AF615321

North American Railroad Stations

Also by Julian Cavalier:

American Castles

Edited With a Foreword by Frederick A. Platt

South Brunswick and New York: A. S. Barnes and Company London: Thomas Yoseloff Ltd

North American Railroad Stations

JULIAN CAVALIER

A. S. Barnes and Co., Inc.
Cranbury, New Jersey 08512

Thomas Yoseloff Ltd
Magdalen House
136-148 Tooley Street
London SE1 2TT, England

Library of Congress Cataloging in Publication Data

Cavalier, Julian, 1931-
North American railroad stations.

Includes index.
1. Railroads—United States—Stations. 2. Railroads—Canada—Stations. I. Title.
TF302.U54C38 385'.314 77-84564
ISBN 0-498-02121-1

PRINTED IN THE UNITED STATES OF AMERICA

Contents

This work is dedicated to the memory of my beloved parents, Helen (1912-1942), and Anthony (1907-1974)

Foreword

Anyone who loves trains must also love train stations, for they are the boarding points to railroading adventure. The city's giant terminal and the countryside's tiny shelter will both arouse the passions of people devoted to the most civilized and pleasurable means of transportation yet devised. Even in those regrettable instances when service along a line has been discontinued, the stations can remain to echo the railroads' glory days.

Railroad stations have been part of train travel almost from the start, but not quite. The first railroad, the Stockton & Darlington Railroad, got under way in England, on September 27, 1825, when thirty-four carriages totaling ninety tons were pulled by *Locomotion No. 1,* invented by George Stephenson. The first railroad station, the still-standing Mt. Clare station, opened in Baltimore, Maryland on May 22, 1830. Since stations quickly came to seem natural to railroading and inseparable from it, does not mean they were always thought of that way. Timetables from the 1830s often announced that the train would leave "from the vicinity of" a certain intersection. Tracks in the period were laid down the center of city streets, and passengers simply would go to that corner and look around for the waiting train. Need for general agreement, as to where the train would stand still to pick up and discharge passengers, gave rise to railroad stations.

The word "station" has the same Latin derivation as the word "stand." Strictly speaking, a "station" is a stopping place along a line, but the word is also used for a "terminal". Terminal means the building at which the tracks come to an end, and it is never applied to a through station. A "depot" is technically a storehouse or warehouse for freight, but since the freight station was not a separate structure but often shares a structure with the passenger station, the term came to signify a railroad station in general.

The emergence of the railroad station was one of the few instances in history when an entirely new type of building had to be created. This is not to imply that stations, in what would become their basic form, appeared full-blown from the minds of the first station designers. They sought precedents for the new kind of building, in several forms already familiar to them.

For example, just as the first passenger cars were stagecoach bodies on railroad wheels, a large number of early stations were fashioned after the inns at which the real stagecoaches had halted en route. The railroad station, in this case, would be built as a hotel and restaurant with the extra feature of a platform at the rear, beside which would run the tracks. Country shelters, that stagecoach transport companies had supplied for their riders, became similarly the open shelters of the whistle stops.

The most abundant type of building was the house, with variations also commonly used as a model for early train stations. Passengers would congregate in what would have been the parlor, while the second story held an apartment for a railroad employee such as the stationmaster. The Greek Revival station might be seen as the first to take on an independent form, but it was actually one more adaptation of the columned temple that at the time was being outfitted as residences, churches, banks, and buildings of every other sort.

Twenty years (1830 until about 1850) were needed for the station to find its own image. Naturally it retained some features of the earlier kinds of buildings from which it had been modeled. From the inn came a variety of services available at many stops, such as the bootblack or saloon. From the house came the wide range of acceptable building materials and the desire to fit in with the neighborhood. The new type of building came in countless shapes, but all drew from a number of features held in common such as a waiting room, ticket office, freight room, platform, rest rooms, and the roof that protected people waiting outside and was quick to become the mark most readily identifying a structure in possession of one as a railroad station. Of course, the open shelter kept to its humble ways, although sometimes even it gained a newsstand.

While the basic form had been achieved, the average antebellum station was never to be more than plain. If it did not have much character, that was because railroading itself did not either. The railroads were barely past the experimental stage since their service was often irregular, their financial resources were slight, and their future network of interconnecting lines was a straggling collection of unrelated roads that had not so much as settled on a standard gauge. Little wonder then that their accompanying stations did not usually have very much personality or attractiveness.

An emphatically unattractive feature of Southern U.S. railroad stations, then and until recently, was that passenger facilities were doubled because of racial discrimination. The plans involved were ingenious though: a pair of waiting rooms would often be separated by a single ticket office, and one kitchen might serve two restaurants.

The Civil War might be argued to have advanced the railroads by demonstrating their abilities at carrying both passengers and freight, in the shape of soldiers and supplies. Development was more likely hindered by the conflict that, for instance, postponed the building of the American transcontinental railroad, planned before the war but not executed until after. In either case, the railroads, like the United States as a whole, entered following the war an enterprising era, in which the stations strove to express the new status of the railroads.

More than fifty years of practically uninterrupted peace, bounded by the Civil War and World War I, allowed the United States, previously an agrarian society, to erupt as an industrial one. Of all the era's industries, the one holding the greatest fascination for the public was the railroads, for they were a subconscious symbol of the growth and energy of the times. A similar period opened for Canada with the formation of the Dominion in 1867. Railroading had special significance for a country that, despite its vast northern territories, is in terms of population a long, narrow strip, like Chile. The building of the transcontinental Canadian Pacific Railway was of prime importance in unifying the new nation and today is studied in Canadian schools, not unlike the War of Independence is in schools of the United States.

Upon proving unequal to the grandeur in which this half-century envisioned itself, architecture—not of railroad stations alone, but of all buildings—underwent a change as well. When the era began, Victorian design still dominated, and the natural move was to lavish more and more ornament on that style. Victorianism soon broke down under the weight of the increased elegance. During the transition period of the 1880s, it was superseded (although it tarried for decades) by traditional styles that represented every architectural period's finest products, to which the new age unhesitatingly felt itself entitled. The era is probably best remembered architecturally for the colossal mansions of its recent millionaires, but each style and size of building benefited, certainly including railroad stations large and small. Traditional styles, in which railroad stations have most often been designed, can be divided into these four groups:

Classical architecture was that of Greece, Rome, and the Renaissance. Ease in handling the ornament upon which the age insisted, made it the era's overall favorite style for every type of building. Often used for the big stations of the larger cities because of its monumentality, it produced handsome structures of all sizes.

English styles include Tudor, the college campus style; its subpart Elizabethan, in which vertical and horizontal timbers frame stucco panels; Jacobean, with its distinctive roof line; and Queen Anne, which in North America became the shingle style. Particularly in the suburbs where such styles were popular, they brought to mind English country life. These four types are descendants of Gothic, a style not exclusively English, that was used for train stations on occasion. Georgian is also an English style widely built by the railroads, but by its nature comes more naturally under the classical category.

Romanesque, a style from 9th- to 12th-century Europe, was fancied by architects whose theories gave them special interest in revealing the construction of a building. The style can be recognized by its massive masonry and round arches.

Mediterranean is based on the buildings surrounding that sea, and is typified by flat walls and rambling form. Warmer areas of the U.S. such as Florida, the Southwest, and southern California were very receptive to this style. Its relation, the Spanish Mission style was also adopted for Western railroad stations. Much earlier, the missionaries, who had pioneered California, wanted to reproduce the Spanish and Mexican buildings they were accustomed to but had to settle for far less grand structures, that could be made by Indian workmen with local materials such as adobe. As it turned out, the resultant style was ideal for railroad stations, because its plan, like that of a station, was one-story and covered an ample area; its arches and arcades were well suited to passenger control; its wide eaves acted as an exterior shelter; its cost, even though stucco had replaced adobe, was low; and often its design even included a bell tower as a match for the eastern stations, where having a tower of some sort was an obsession.

Some of the era's architects, like Bradford L. Gilbert, went so far as to specialize in the design of railroad stations. Sad to say, most stations have long ago become dissociated from the names of their architects, even those who were not railroad employees and so did not sign their blueprints anonymously as "Architect of the ——— Railroad." The exact years of design and construction have often been lost also.

It is certain though, that the railroad station by now had come to be one of the most important buildings in the community. Rural settlements saw the station as their primary link with the rest of the world. The station offered instant communication and news through its telegraph wire, and brought the thrill of a traveling circus or a one-night-stand theatrical. Indeed, practically everyone and everything arriving in town from much distance came by way of the railroad station. Maintenance, befitting the station's importance, was painstaking, not only in the building itself, but also in the groundskeeping, consisting of manicured lawns, raked pebble paths, and sometimes even the railroad's emblem in flowers.

The era admired, rather than feared, its cities, so they erected immense railroad stations there and called them the Gateways to the Cities. Referring to the gateways in the walls surrounding ancient cities, the term is appropriate, because both were where travelers gained access. The old gateways were often the scene of bazaars, and similarly the big stations not only held a variety of shops, but also made provisions for cabstands, stationmaster, baggage, mail, express, announcer, information, general waiting room, parcel check, newsstand, lunch room, restaurant, telephone, telegraph, rest rooms, bootblack, barber shop, and beauty parlor. Like the ancient gateway, the city station was given a masterful design to impress the grandeur of the city upon whomever was entering it.

Style was not alone in altering the appearance of the railroad station. Toward the turn of the century, safety had begun to dictate that grade crossings or the intersecting of tracks and streets at the same level be eliminated, except in the least populated areas. Much deliberation by architects, engineers, and local governments was called for, especially in suburban towns where the shopping district, the center of the community, gravitated to the train station. (Incidentally, suburban stations were often pairs of buildings, a larger one on the inbound track, a smaller one on the outbound.) If the track level could not be significantly changed, a decision would have to be made as to which would look and work better, a humpbacked over bridge or a steeply inclined tunnel under the tracks. If the track level was to be raised or lowered, the town fathers were left with the prospect of either an elevated line or a ditch through the center of the community. Fortunately the state of design at the time was strong enough to allow for agreeable solutions.

Despite the recurring second-story apartment for

a railroad employee, the station had been essentially a one-level operation. Though the station building itself usually remained so, a change was effected in the layout of the grounds. When the track level was unaltered, the station and its surroundings would naturally stay the same, with the added presence of a bridge or tunnel, neither of which did much for the beauty of the area, but could be fashioned to be almost overlooked. When the tracks were raised, a new station would commonly be built to sit at the new height, atop a man-made hill with landscaped slopes and ascending pathways. When the tracks were in a ditch (which generally proved the most aesthetic of these several possibilities), the train station would similarly be sunken with stairs running down to it, although there were instances of the ticket office and one waiting room being at street level, while the platform and a second waiting room were at the lower level.

Multiple levels were also used for the big stations of the cities. Reading Terminal, built in Philadelphia between 1892 and 1893 and still standing is a giant example of a raised one-level plan. The ground floor holds stores and the famed Reading Terminal Market, while the passenger station, reached by stairs and elevators originally and now by escalators too, is on the main floor just above. At the front of the station, the head house, designed by Francis Kimball, holds in its upper stories the main offices of the Reading Railroad. Completed in 1913 for the New York Central, which like the Reading is now part of Conrail, Grand Central Terminal, whose fantastic engineering feats were accomplished by William G. Wilgus and the firm of Reed and Stem, brought trains in on two levels. The lower one is used for commuter runs and the upper one for long-distance travel. Passenger circulation through the station was expedited by a system of well placed ramps.

Another alteration, for safety's sake, was necessitated by passengers crossing the tracks at suburban stations. Occasionally the distance around was short enough that the sidewalks of the vehicle road would be used, but at many stations, due to their size or setting, an overhead walkway or an underpass was needed. The overhead walkway, although some were gracefully arched, looked ungainly, while the underpass had the drainage and other problems of subterranean construction, so again a difficult decision had to be made. Still one more change in some stations, at about this time, was the adoption of high platforms that came up to the floor level of the passenger car, so that riders could step directly from the platform into the vestibule without using the car's stairs.

The high platform had earlier come into use at many freight facilities, because it made the loading of boxcars easier and faster. Sometimes the platform would be wide enough to reach the car door, and sometimes there would be an intervening space to be filled in by a baggage truck. The freight depot itself, when built apart from the passenger station, was usually close by and similar, in style, to it, but simpler and stripped of conveniences needed for travelers.

By the 1920s, the railroads had become upstanding corporations. The manipulators, who had presided over the railroads' adolescence and often plundered them, had been replaced by businessmen, who managed them soundly and efficiently. Railroad stations had become solid citizens, but also rather set in their ways. From the suburban stations in what had been reduced to "pseudor-Tudor," as the modern architect Eero Saarinen called the style, to the ponderous classical stations for the cities such as Chicago's Union Station, designed in 1926 by Graham, Anderson, Probst & White, new stations were staid versions of their pre-World War I ancestors. Even the Spanish Mission style, in which simplicity was inherent to the design, had increased in stateliness, as witnesses the large Union Passenger Terminal, built at Los Angeles, from 1934 to 1939, to the designs of Donald and John Parkinson. While these buildings are somewhat lacking in creativeness, it must nonetheless be said that they are handsome and imposing.

Powering trains by electricity, a practice that grew rapidly in the 1930s, greatly cut restrictions on architects. The earliest designers of stations had been especially fond of running the trains through the station buildings, often of wood and therefore short-lived due to sparks. It was soon found preferable, in most cases, to leave the trains in the open and cover the platform with an overhanging roof or a canopy. Taking a fired-up steam locomotive indoors, when that was desired, called for a very high roof, like those of the train sheds. These dramatic structures with their vast vaults were once a part of many terminals and through stations, but are now nearly extinct. A survivor is the train shed designed by the Wilson Brothers for Reading Terminal. Electrification made it possible for trains to pass beneath low ceilings, allowing much greater flexibility in station design.

Such freedom meant, however, that the huge station buildings of the cities, where land is at a premium, could be done away with and replaced by structures thought to be more profitable. Not only could the tracks be placed underground (if they were not there already), but so could the rest of the station. Following the Second World War, the quick and thorough transition from trains to planes began, and by the early 1950s the railroads started demolishing city stations. In the suburbs and country, stations economically unworthy even of being torn down were abandoned hastily and in quantity, while many of those kept in service were allowed to deteriorate.

The destruction continued pretty much unchecked until the announcement in the 1960s that Pennsylvania Station in New York City was to be torn down. Designed in 1906 by Charles Follen McKim of that era's most prominent architectural firm, McKim, Mead and White, the station had been occasioned by the new tunnel that the Pennsylvania Railroad had built beneath the Hudson River so that the PRR could have direct access to Manhattan. Both outside and in, the classical edifice was one of the most beautiful buildings of any kind in North America. The public was outraged at the prospect of the station's disappearance, and even architects, usually the last group to be concerned about a demolition, worked to prevent it. All efforts failed, and on the site arose the new Madison Square Garden sports arena with a new, sterile station beneath it. However, the episode sparked the movement for architectural preservation. People realized that under prevailing conditions in architecture, whatever was torn down would almost inevitably be replaced by something worse. New York's other monumental station, Grand Central Terminal, is now threatened. Its splendid engineering, already mentioned, is equalled by its Beaux Arts design created in 1911 by Whitney Warren, of Warren and Wetmore. Grand Central too rates as one of the finest buildings ever erected on this continent. How well the lesson of Penn Station has been learned remains to be seen.

Saving a giant station is a giant labor, best undertaken by a committee, but preserving a smaller one also requires gumption. Nevertheless it is being increasingly done throughout North America, and offers great rewards to whomever attempts it. Civic groups and even high school classes sponsor projects in which members paint and otherwise renovate a local station that is still serving its original purpose; a garden club takes charge of the grounds. For their own needs do individuals, organizations, and businesses remodel unused stations into stores, art galleries, dental offices, factories, private homes—practically any purpose requiring a small building. (Some advice: When altering a railroad station to some other use, it is best to keep at least the exterior looking like a railroad station, since that, after all, is what was set out to be saved. When renovating working stations, it is similarly wise to avoid mod colors and decorations that would not have been chosen originally.) A sturdy and architecturally appealing building can be bought and refurbished for usually far less than the cost of erecting a new, bland structure. The extra advantage to saving a station is that part of the excitement and grand style of railroading will remain as proof to the future that technology at its best considers the human dimension always.

In publishing as in preservation, the largest stations have regularly been more conspicuous, but thanks to this collection of photographs compiled and captioned by Julian Cavalier, smaller stations are being compensated for this neglect. May this book serve as a reminder of the compact treasures that have been lost and as a goad to the preservation of those that still survive.

Frederick Platt

Preface

This work is offered as a limited pictorial review with some informative comment for each structure presented that have locations throughout both the United States mainland and Canada. The main purpose of the work is to show pictorially some of the great number of various railroad station designs that still exist, and a few that have long since vanished. Emphasis is placed on the smaller and medium sized buildings as they are fast disappearing from the railroad scene at an ever increasing rate.

Many of the photographs presented have been obtained from private sources and collections of individuals who have wisely taken photographs of railroad stations with the knowledge or foresight that such buildings may soon pass, like the steam locomotive, into railroad history. Several railroads have also made some photo views available from their files, and numerous other sources were sought for available views.

The photographs are arranged in alphabetical order for easy reference in conjunction with the index listing. It should be noted that many of the original railroads mentioned in this work no longer exist independently, have been absorbed, or merged with other, usually larger railroads, even though their rail lines still exist and continue to be used. All stations are listed by their names, which in almost every case is the same as the geographical name place in which they are located. The term, "station" is used here and in the Introduction to mean a railroad depot or station building, in general. It is hoped that the buildings pictured here may be viewed as a small pictorial record, a sampling of railroad station architecture of which the majority viewed here still survive at this writing, but may not in the near future.

Acknowledgements

I wish to express my thanks and appreciation to the many individuals whose photographs are presented in this work, to the railroads represented here, the historical societies, and other agencies, all of whom have helped in making their material available during my research on this project. Special thanks to George B. MacKay and Henry E. Bender, both of whom searched and made their negatives available from their collections; and to William F. Rapp, Jr., of the Railroad Station Historical Society, Crete, Nebraska, who made much data available and searched through his large collection for many of the photographs presented. Also my thanks to Herbert H. Harwood, Jr.; George Kraus; Stan F. Styles; John Uckley; Howard W. Ameling; and all others whose names appear in the photo credits of this work.

Introduction

When passenger rail travel first began in North America there were no station facilities along the lines as we know them today. Departure points then were at various locations such as at a known crossroads, hotels, and at other prominant places in a town or other locations. Tickets were usually sold by the conductor, from certain local places of business, and at departure points, but in all cases there was no station facility for passenger convenience, as later developed in the early 1840s with the growth of the industry. With the expansion of those existing railroads, and the development of new railroads, more and more passenger stations were built until they numbered in the thousands.

With the increased development of the railroad station, their designs, sizes, and facilities also grew more elaborately. Railroad station architecture is most often of a unique character, though varied standards of designs used by many railroads had also evolved, often designed and built by the railroad using their own engineering departments and company forces. The distinctive forms of design and style continued to emerge along with standard designs, most of which also varied in some form or other to suit local conditions at the site. It was not uncommon to see even very unique designs, one of a kind, duplicated in no other place.

Often, as the railroads built their lines, the engineers plotted communities. Often the first building to be built was the railroad station, usually at the edge of a new town, while the town's first real estate agent began setting up shop nearby. Frequently, the station building, in those early years, housed not only the agent and his family, but also a hotel and restaurant, bringing new amenities to an otherwise inhospitable land. The station thus became the most important building in the community because it not only offered such amenities, but it also was the doorway to an isolated world. It offered news and the means of instant communication with the rest of the world through its telegraph lines, it offered excitement as trains discharged passengers or brought in a traveling circus or a troop of one-night-stand actors, and it was the source of outgoing and incoming mail, merchandise, and other products of trade and commerce.

In the beginning of the 1920s, rail passenger service had already begun to decline. Passenger service had become unprofitable to the railroads for many years especially since the 1930s. For a brief period, during the World War II years, business increased, but soon afterwards the pre-1940 period of decline resumed which continued until the present time where such service has become drastically limited.

As with the decline of rail passenger service, the railroad station also began the process of passing out of existence. It is not uncommon now for dozens of railroad stations to close each month, to have doors and windows boarded up and become abandoned. More often than not closed stations soon become the victim of wrecking crews, and with their disappearance a little part of railroad station architecture adds to the many already passed from view. In an ever changing world, perhaps a sense of stability also passes with them. The vast scale of station demolitions continues at this time, almost like a vanishing species of something once familiar and plentiful in communities, and taken for granted as part of a stability of progress.

It can be viewed now with reasonable certainty and with the evidence at hand of station closings and demolitions, that the currently existing stations are in a stage of progressing extinction, and especially so, more recently, since the advent of Amtrak in the United States. Like the steam locomotive, the railroad stations as we knew them are becoming a thing of the past as they will for the greater majority of stations that manage to survive still.

Fortunately, many stations are being reused for other nonrailroad purposes so that at least some small portion of them have escaped the wrecking crews. It is hoped that many more stations can be reused, even if only to serve again in some other useful capacity that would at least help to retain and preserve some examples of early railroad station architecture, of designs that have long since ceased to be built. Some of the stations pictured here may appear as common, and as such, thought plentiful so that interest in their designs and styles of architecture may be of slight importance to some. It perhaps should be recognized that it is this style that is fast disappearing and are no longer plentiful, as might otherwise be assumed. Regardless of the architectural style and criticisms of past designs that could be observed, all styles and designs are of interest to the student of railroad station architecture or to anyone who has a genuine interest in such railroad buildings. The majority of railroad stations presented here still exist at this writing. A considerable number of them have also been closed and fallen into disrepair since these photo views were taken, while still others continue to survive mainly for freight only service. Very few pre-Civil War stations survive, and those built since then to the turn of the present century are fast becoming scarce due to continued closings and demolitions.

It is difficult to categorize North American railroad stations because of the great variety in styles and designs, though some attempt has been made in the past to separate them by locations and sizes. One major category used by the railroads, in earlier times, was the standard type station that had many variations in designs and styles. Such standard types also varied with each of the railroads, that had such standards of design that included stations of one room to those having many rooms, as well as styles that included accomodations for living quarters as well as combination passenger and freight designs. The on the site erection of such so-called standard designs, more often than not, required alterations to suit local conditions. During the course of the station's service further alterations were common, either for additions to the building because of increasing business, or alterations due to the decline of passenger business such as stations, converted for freight only service. Thus the term, standard design, is an ambiguous one and might be considered in a broad sense of the word.

The railroad station is perhaps one of the most interesting styles of architecture built for a specific purpose and on a large scale in North America. Though the designs and sizes do vary considerably, their character is distinctive and is a main link that distinguishes them from all other forms of architecture in this respect. Whether they are small, wood frame, flag stop stations or large elaborate buildings of marble, this character of purpose and design is retained. Almost every style of architecture has been used over the years with some noted design characteristics that may group certain stations as to their architectural styles though this also involves considerable design variations within such groups. These may be referred to loosely as the Victorian, Romanesque, Spanish Revival, Mediterranian, Georgian, California Mission Revival, etc., and include even the variations of modern designs and styles of station architecture, especially those built after World War II.

Some mention might be made here of the large variety of building materials used in the basic construction. Perhaps the most common materials used for the majority of stations are the wood frame construction and the brick constructed stations, with numerous variations combining the two and with other building materials. With most of the Spanish Revival, Mediterranian, and California Mission Revival styles, the basic building material for its walls is brick, covered over with a stucco finish. With such styles the use of the red Spanish tiled roofing was generally incorporated as a part of the overall design style. With wood frame, the styles and design variations are enormous. Other materials have been used to a great extent such as natural and cut stone, flag stone, wood frame with stucco finish, brick and stone, brick and stucco, log construction, log and stone, and so on, as such a listing of variety is considerable. In every case, however, the character of the building as a railroad station is present, regardless of the building materials used.

It is well known to the student of railroading, or those who persue the subject, that the current large scale demolition of stations in North America is an established fact. However, it is not as well known to the public in general, that some small portion of existing stations are being reused for other nonrailroad service. In recent years many stations have been leased or purchased outright, and converted to other practical uses. In most cases the exterior character of the station building has been retained and in so doing, some measure of preservation is retained for various styles of station designs. In other instances, complete restoration has been made to the buildings so as to fully retain its original character. Regardless of the reason of reuse, such stations do exist as functioning buildings of practical value that at the same time retains some forms of early station architecture that are no longer being built. Whether they be retained for historical reasons or as practical reusable buildings, their architectural values may be appreciated more as time passes.

Though a substantial number of stations still exist in North America, their continued demolition is not seen as a great threat to the majority of the general public, most of whom could not care one way or the other. But for those who have some foresight to the real situation, it is evident that the need to document and record these buildings must be done now. It has been estimated that only about one-fifth of the stations still exist, from the number estimated to have existed at the turn of the present century in North America. In the United States this estimated figure for the entire country is approximately twenty thousand. Though this figure may seem still considerable the monthly rate of closings, in just the last recent few years, is equally considerable. An example of station closings in the Province of Saskatchewan, Canada, in 1975 by one railroad and during one month, was fifty-five stations. This is not always the case but it does illustrate the determination of the railroads to close buildings enmass when the situation calls for it. Over the entire country the number of closings is, of course, much greater overall, not to mention all those being closed on the United States mainland each month. It becomes more evident now that, on the whole, a large number of stations are closing each month within these two countries. It would not seem unreasonable to assume then that the estimated number of existing stations is not a large one, taking into consideration the excellerated rate of closings and demolitions that have taken place and continue to take place since the early 1970s.

The appreciation of railway station architecture can perhaps be better realized on a comparison of those who have an equal appreciation of the steam locomotive. As the decline of the railroad station continues an awareness and interest in them becomes more acute, especially for those who have already had an interest in the subject.

This work is not intended, nor does it pretend, to be a complete work in showing all the various designs that have been built. It is hoped that this modest work will be of interest to the reader, in illustrating some varieties of railroad station architecture that have been built in North America in the past, by a variety of railroads.

North American Railroad Stations

Anchorage, Alaska **Alaska Railroad**

This March 1918 view shows the activity at the old Anchorage depot caused by the departure of young men to join the army, during World War I. The two-story, wood-frame depot stood near the shores of Cook Inlet, on the Anchorage-Seward line.
Courtesy H. A. Haag, Alaska Railroad

Fairbanks, Alaska **Alaska Railroad**

The old, wood-frame station at Fairbanks was built in 1923 on the Mt. McKinley Park Route just south of the Yukon River. This view shows the station in a snowy scene, during the winter of 1947-1948. The station served until 1961 when it was replaced by another depot.
Alaska Railroad

FAIRBANKS

Nenana, Alaska **Alaska Railroad**

A distant, trackside view of the Nenana station taken in 1951. The station is currently still in service to passenger and mixed trains. It is located 297 miles north of Anchorage and 59 miles south of Fairbanks, on the Alaska Railroad's Mt. McKinley Park Route.
Alaska Railroad

Nenana, Alaska **Alaska Railroad**

Nenana station, built in 1922, is a wood-frame, combination (passenger-and-freight) station consisting of a main block of two stories with a single-story wing. The first floor is shingled, and the upper covered with horizontal siding. This 1951 view is of the rear of the building.
Alaska Railroad

Seward, Alaska **Alaska Railroad**

Seward station as it looked on October 27, 1948. The station is believed to have been originally built by the Alaska Central, and acquired by the Alaska Railroad in June 1917. It was moved in August 1928, and its ownership transferred to the City of Seward; it thereafter was used as a ferry terminal.
Alaska Railroad

Samaria, Michigan **Ann Arbor Railroad**

Samaria station was built in the late 1880s and located in the small village of Samaria, on the main line of the Ann Arbor Railroad between Toledo, Ohio, and Ann Arbor, Michigan. The photograph is a rare view, for the station was demolished many years ago. First trains operated on the line from Toledo to Ann Arbor on May 16, 1879 with celebrations and a special train.
Collection of Everett J. Payette

Samaria, Michigan **Ann Arbor Railroad**

About 1911, the Ann Arbor Railroad placed several gasoline-operated cars on passenger runs along its line, stopping them at every highway crossing, village, and city, to accommodate passengers and take on farm produce and milk cans. The service ended during the 1930s. The photograph, dating from around 1911, shows one such motor car at the Samaria station.
Collection of Everett J. Payette

Aguila, Arizona **Atchison, Topeka & Santa Fe**

The Aguila depot is located 22.2 miles west of Matthie, northwest of Phoenix, Arizona, on the railroad's Matthie, Parker, and Cadiz line. The depot is of conventional design, with gabled roof and open freight platform at the freight end of the building. When this view of August 26, 1967 was taken, the depot was still in use, but for freight service only. William F. Rapp, Jr.

Arcadia, California **Atchison, Topeka & Santa Fe**

The Arcadia station was built in 1895 and located on the railroad's line from Albuquerque to Los Angeles. The wood-frame building was dismantled in large sections and moved to the site of the Southern California Chapter of the Railway and Locomotive Historical Society, Inc., where it was reassembled and now stands. After relocation, the building was repaired, restored, and repainted. Features of the edifice are the multilevel roof, the towers, and the extensive ornamentation on the exterior. This view of July 17, 1967 shows the station at its original site. Henry E. Bender

Fort Madison, Iowa **Atchison, Topeka & Santa Fe**

Fort Madison station is on the Santa Fe's main line, 232.4 miles southwest of Chicago. As solid as the Midwest, this brick, Tudor station with its massive three-story tower replaced the original wood-frame depot. The newer station was still in active service when this photograph was taken in July 1962.
William F. Rapp, Jr.

Grand Canyon, Arizona **Atchison, Topeka & Santa Fe**

A special design was needed for this station in a special place, and so a large, log structure was constructed. Completed in 1909 and located 90 miles southwest of Flagstaff, it was a terminal station on the south rim of the Grand Canyon until 1968, when regular service ended. The building was still in good condition when this view was made on May 1, 1969, but it has since been vacant and fallen into disrepair.
Henry E. Bender, Jr.

Hardin, Missouri **Atchison, Topeka & Santa Fe**

Adding a two-story section to an earlier one-story depot produced the unusual design of this wood-frame station. Although it handles passengers, the depot is used mainly for freight. The photograph was made October 16, 1972.
William F. Rapp, Jr.

Las Cruces, New Mexico **Atchison, Topeka & Santa Fe**

Las Cruces is a medium-sized, Mediterranean style station on the Santa Fe's Albuquerque and El Paso line, 210 miles south of Albuquerque. Adaptations of the railroad's emblem crown the gables. The station has passenger and freight facilities, and rail-auto service has been available. As seen in this view of June 20, 1966, the building is still in good condition.
Van W. Best; W. F. Rapp Collection

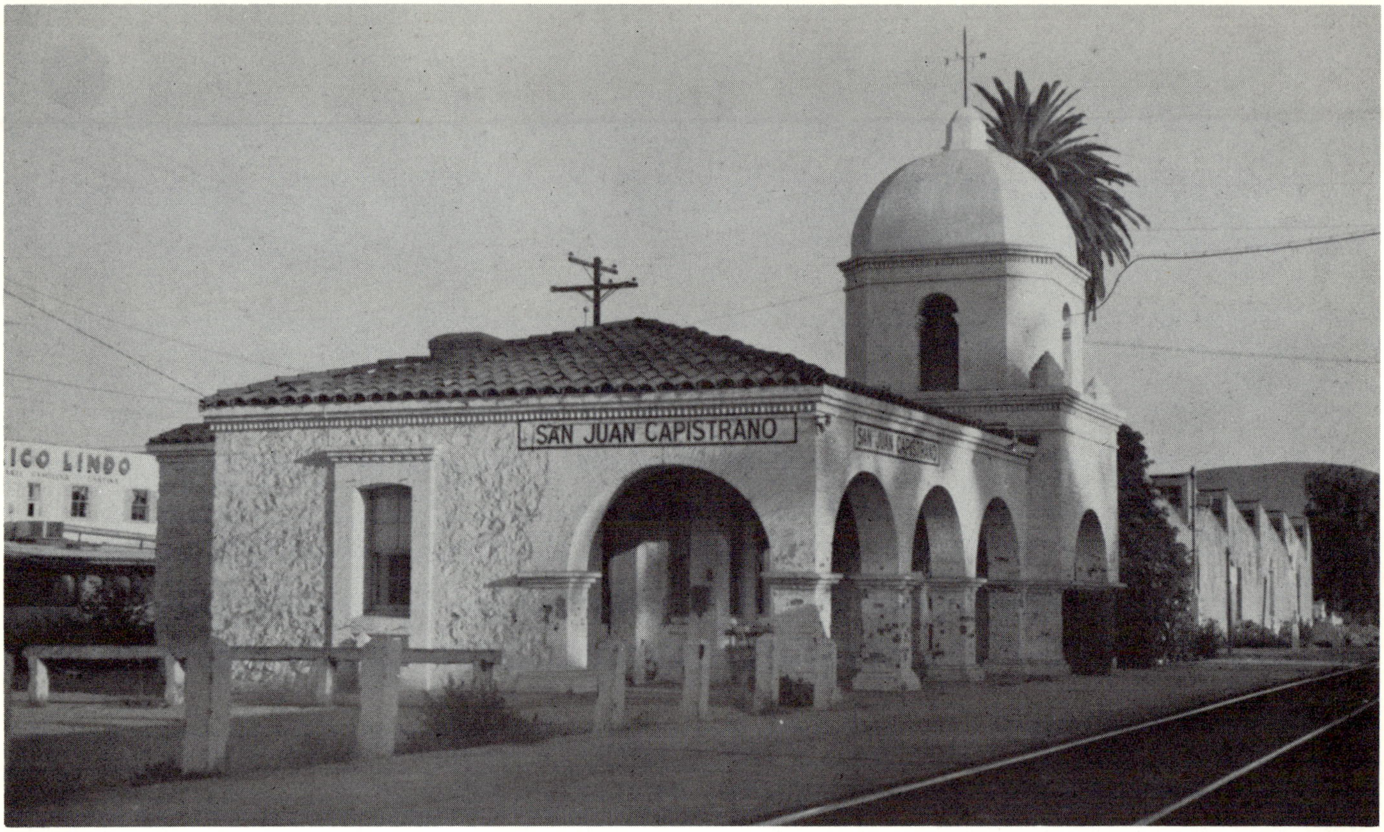

ICO LINDO
SAN JUAN CAPISTRANO

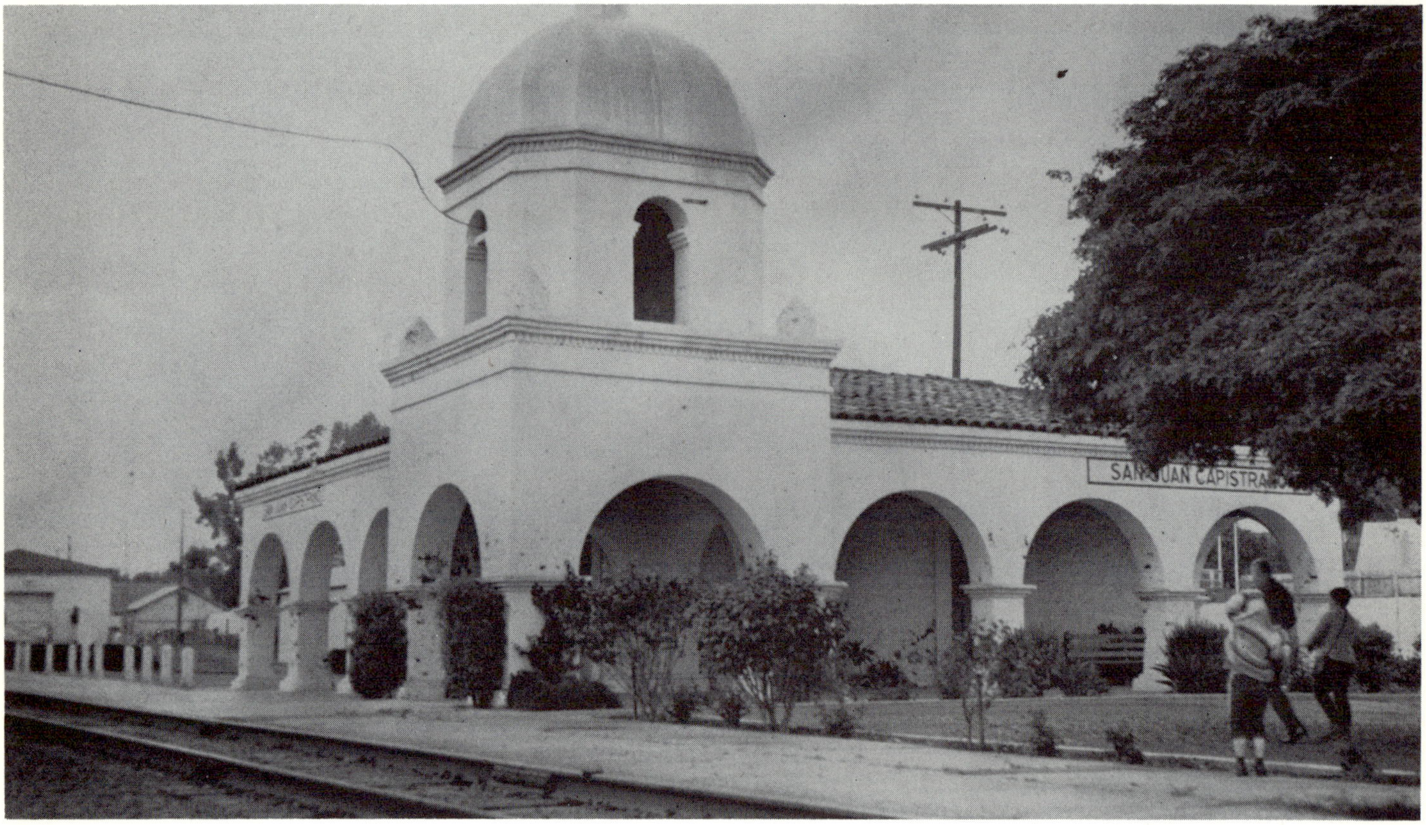

SAN JUAN CAPISTRANO

San Juan Capistrano, California
Atchison, Topeka & Santa Fe

As originally built in 1895, this station contained three rooms: an office, a waiting room, and a baggage room. The open colonnade runs along two sides of the Spanish Mission structure, which is mainly of brick with a red tile roof.
Henry E. Bender, Jr.

Seligman, Arizona **Atchison, Topeka & Santa Fe**

Of handsome design and proportions, Seligman station is located in northwestern Arizona, 84 miles west of Flagstaff on the Santa Fe main line south of the Grand Canyon. The large, two-story, Elizabethan building is of brick with the upstairs walls finished in stucco and timber, and topped by a red tile roof. Recessed into the first floor, at trackside, is a loggia. Still in active service, the station is in excellent condition as seen in this view of May 1, 1969.
Henry E. Bender, Jr.

San Juan Capistrano, California
Atchison, Topeka & Santa Fe

Located 57.6 miles south of Los Angeles, the station is on the Santa Fe's surf line. About 200 feet from one end of the building is the freight house, seen at the extreme left of this view. The railroad has leased the station to the Capistrano Depot Restaurant, which opened for business, in the building, in February 1975. The station is shown in this April 1967 view as it looked prior to the conversion.
Stan F. Styles

Waynoka, Oklahoma **Atchison, Topeka & Santa Fe**

The station sits adjacent to the former Harvey House (not shown), both on the east side of the Santa Fe main line. The brick Tudor station sporting stone versions of the Santa Fe emblem serves passengers no longer, but is still used for freight. Both structures were built in 1910, visited in 1929 by Charles A. Lindbergh, and placed on the National Register of Historic Places in 1974. The Harvey House, which served the aviator a meal, is now a reading room and dormitory for crewmen laying over between runs.
Santa Fe Railway

Wickenburg, Arizona **Atchison, Topeka & Santa Fe**

Wickenburg, which the sign clearly identifies, is a neat and trim combination depot on the Ash Fork and Phoenix line. Baggage trucks alongside and the air conditioner on the roof at right indicate the depot was very much in service when this view was taken on June 19, 1966.
William F. Rapp, Jr.

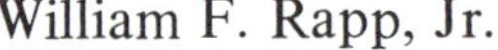

Fayetteville, North Carolina **Atlantic Coast Line**

Fayetteville station was built in 1912 as a passenger station by the Atlantic Coast Line Railroad, later Seaboard Coast Line. This view of January 19, 1975 shows the station still in service by AMTRAK. The building is basically of brick construction with a Gambrel style roof design, a dominating feature of the station. Amtrak occupies a portion of the station but a larger portion is still utilized by the Seaboard Coast Line Railroad for office space and freight agency.
R.D. Patton Railroad Photo.

Brunswick, Maryland **Baltimore & Ohio**

This wood-frame station combines admirably the informality of shingled walls with the grace of classical elements, most evident on the dormers, where are found both straight and scalloped shingles as well as small Palladian windows. The station is seen in a September 3, 1972 view.
R. D. Patton

Brunswick, Maryland **Baltimore & Ohio**

On the Baltimore & Ohio main line, 49 miles west of Baltimore, this station is still in service. It also houses a railway express agency. The prominent sign on the long wall in this picture states that the station is for westbound trains only. The station is well-maintained, as proven by this photograph from April 1975.
Wilbur C. Thurman Collection

Dickerson, Maryland **Baltimore & Ohio**

One of the B&O's standard designs in the late-19th century, this charming, little station is on the Washington-Cumberland line, 35 miles west of Washington, D.C. This wood-frame structure was erected in 1871, a date displayed along with the railroad's initials in a pediment. The trackside bay is triangular. Still surviving, the station is seen in a photograph made in January 1975.
Wilbur C. Thurman Collection

ott City, Maryland **Baltimore & Ohio**

Baltimore & Ohio Railroad still maintains some of the st railroad stations in the United States such as the one at ott City which was built during a period between 0-1831 and is still in service today. The cut stone building changed very little since it was built. It is on two levels, second floor being at track level and is used today as a way Express office, and is open to the public. This view ws the depot as it looked in June, 1970.
bert H. Harwood, Jr.

Huntington, West Virginia **Baltimore & Ohio**

Built in 1887, the two-story, brick station has been used jointly by the Baltimore & Ohio and the Chesapeake & Ohio railroads. This picture shows a train westbound to Kenova. Although no longer used by either railroad today, the station seen here in a May 1956 photograph has been entered in the National Register of Historic Places.
Herbert H. Harwood, Jr.

Laurel, Maryland **Baltimore & Ohio**

Laurel station was built in 1884 from the plans of E. Francis Baldwin, who also designed the Maryland stations at Point-of-Rocks and Mt. Royal. The building, which is on the National Register of Historic Places, is still in reasonably good condition, as witnessed by this view of January 1970. Herbert H. Harwood, Jr.

Midland City, Ohio **Baltimore & Ohio**

Located 43 miles east of Cincinnati, the old Midland City station is where the B&O's Cincinnati-Washington line meets its Midland City-Pittsburgh line. The two-story, wood-frame building is set on a brick platform between the two lines. As seen in this August 1967 view, the station is in need of repairs.
Max Miller

Mount Clare, Baltimore, Maryland **Baltimore & Ohio**

Built in 1830, the Mount Clare station is adjacent to the Baltimore & Ohio's great Mount Clare shops, on Poppleton Street just south of Pratt Street. Credited with being the first railroad station in the world, it is excellently preserved. In the background can be seen the huge roundhouse that, together with the station, now holds the railroad museum of the B&O.

Baltimore & Ohio Railroad

Mount Clare, Baltimore, Maryland **Baltimore & Ohio**

It was at Mount Clare station that the first train tickets were sold for trips over the B&O's 13-mile route between Baltimore and Endicott's Mills in 1830. Primitive steam locomotives like the *Tom Thumb* and the *Atlantic* hauled the little cars over tracks that terminated at the railroad station. The paint color has been changed in this present-day photograph from that of the previous picture, taken earlier. Baltimore & Ohio Railroad

Salem, West Virginia **Baltimore & Ohio**

Located on the B&O main line in the northern part of the state, 283 miles west of Washington, D.C., the brick building is enlivened by the fine fenestration that includes a large center window, and by the tile roof with its gable brackets. When this photograph was made in June 1974, the station remained in active service.
David P. Oroszi

erwood, Ohio **Baltimore & Ohio**

is is the very first depot built in Sherwood and was located proximately 71 miles south of Jackson, Michigan. The O main line was a single track when this photo was taken out 1905, the line being built in 1875. This Sherwood pot has long since gone and the interlocking tower in the ckground was demolished in 1970. The depot was also erred to as the Sherwood "Union Depot" at that time.
llection of Everett J. Payette

Intervale, New Hampshire **Boston & Maine**

Near the Main state line, Intervale depot is 61.4 miles northwest of Portland, Maine. Used jointly with the Maine Central, the building was a terminal for the Boston & Maine. The small, wood-frame depot with tall windows is of standard design. The depot was in active service and well maintained when this 1938 view was taken.
Norton D. Clark

Powwow River, New Hampshire **Boston & Maine**

This scene was typical at the small station at Powwow River when a passenger train arrived there, as seen in this view from 1929. The building is simplicity itself: a clapboard block with a hip roof. Of special interest are the old pole lamps and the lamps on the station, as well as those on the rear of the train. Norton D. Clark

Allandale, Ontario **Canadian National Railways**

Originally part of the Toronto, Simcoe & Huron Railway, but now belonging to the Canadian National Railways, the earliest section of this long, rambling station, at Allandale, was built in 1853. This view, made in the late 1930s, shows the old, Italianate division of the building.
Canadian National Railways

Allandale, Ontario **Canadian National Railways**

This photograph shows a portion of the station's flower garden in the foreground. Immediately behind it is the brick addition to the station, and beyond that is the older, wooden structure. The addition was also Italian in inspiration, but the interpretation of a later day than that of the section seen in the previous photograph. The tile roof helps unify the whole station.
Canadian National Railways

Amherst, Nova Scotia **Canadian National Railways**

Amherst station was built by the Intercolonial Railway, which became part of the Canadian National in the 1920s. The station handled both passenger and freight, the freight room being seen at the extreme right in this 1912 picture.
Cavalier Collection

Amherst, Nova Scotia **Canadian National Railways**

The two-story, cut stone station at Amherst is still in active service on the Canadian National Railways. This view, taken June 19, 1970, shows that the station has not altered much since it was built, except that it has been reroofed and the paint color has been changed. The station features broad arched and rectangular groups of windows that are divided vertically, and a single-story freight section.
George B. MacKay

Amqui, Quebec **Canadian National Railways**

South of the St. Lawrence River and 250.3 miles from Gaspe, Quebec, this wood-frame building has second-story windows that extend upward to the dormers. The roof brackets are the same shape at both levels, but a larger size below. This photograph was made on June 12, 1970.
George B. MacKay

Apohaqui, New Brunswick **Canadian National Railways**

This handsome, little station at Apohaqui is located about 50 miles from the city of Moncton. The wood-frame structure has a corner bay, and walls covered by clapboards below and shingles above. In this view of June 23, 1970, the storm windows are still in place.
George B. MacKay

Barachois, Quebec **Canadian National Railways**

On the shore of the Gulf of St. Lawrence, which is seen at right in this view, Barachois station is 25 miles from Gaspe, Quebec. The corners of the first floor are indented below the second, with the roof extending out even further at trackside as a platform shelter. The bilingual sign to the left of that bearing the station name prohibits vehicles from using the platform. The station is well-kept and still in service, as seen in this view of June 13, 1970.
George B. MacKay

Battleford, Saskatchewan **Canadian National Railways**

Built in 1905 by the Canadian Northern Railway, which has since become part of the Canadian National Railways, Battleford station is of a standard design. The stucco building is met at trackside by a wood-plank platform and has living quarters on the second floor. The station was still in use when this view of August 1972 was taken.
Stan F. Styles

Blue River, British Columbia **Canadian National Railways**

Canadian Northern Pacific Railway original drawings indicate that this station was built in 1915. Blue River station, now part of Canadian National Railways, is one of the variations the CNP made on a standard design. This view shows the station as it appeared in August 1972.
Stan F. Styles

Boston Bar, British Columbia Canadian National Railways

To this depot of stucco exterior has been added the wood section seen at right, that probably was used by the station restaurant. Note the wood-plank platform. This view was taken May 17, 1952.
Stan F. Styles

Boston Bar, British Columbia Canadian National Railways

A freight train headed by No. 2509 is seen at Boston Bar station in Fraser Canyon, in this scene taken July 20, 1954. The track inspector has the main line for a short distance.
Stan F. Styles

Brampton, Ontario **Canadian National Railways**
The round towers that flank the rear entrance are seen in this view, as is the arched, open-air waiting space at the southern end of the building. Between stone foundation and slate roof are buff bricks; years of weathering have made their color difficult to detect, although the station is still in excellent condition.
Julian Cavalier

rampton, Ontario **Canadian National Railways**

his station is on the Halton subdivision (the old Stratford ivision), which was originally the Toronto & Guelph ailway, that had incorporated on August 30, 1851 and nalgamated with the Grand Trunk Railway Company of anada on July 1, 1893. The line opened for traffic on July , 1856, the year Brampton station was built. Both line and ation were acquired by Canadian National Railways upon ormation in the 1920s. The station is now a stop on the new O Train commuter run, that has the installation of the nall, modern style shelters and the chain link fence seen in is view of April 27, 1974.
ulian Cavalier

Brighton, Ontario **Canadian National Railways**

Located east of Toronto, Brighton depot is on the Canadian National's Toronto-Montreal main line along Lake Ontario. This station is basically very similar to the Grand Trunk Western depots built in 1859 at New Haven and Richmond, both in Michigan. This brick depot with the flat relief, also in brick, above the arched doorways and around the circular gable windows is seen in a view taken June 16, 1972.
Howard W. Ameling

Dauphin, Manitoba **Canadian National Railways**

Located 176 miles northwest of Winnipeg, this large, two-and-a-half-story, brick station with stone base and trimmings is, though smaller, similar in style to the Canadian National's station at Port Arthur, Ontario. As pictured in this August 1968 view, the station is well-maintained and in active service.
Stan F. Styles

Edson, Alberta **Canadian National Railways**

Edson station is on the Winnipeg-Vancouver main line, 130 miles west of Edmonton. Despite its considerable size, this stucco building is unimpressive overall, though it has some strong features such as the chimneys and the bands surrounding the windows and the lower part of the first story. The station was still in active service when this view was made in August 1967.
Stan F. Styles

Fort Langley, British Columbia

Canadian National Railways

By March 1971 when this photograph was taken, the exterior appearance of the station, so spruced up in the previous picture, had somewhat deteriorated. The two-story, wood-frame building is of standard design with living quarters above. The station is the point stop to Fort Langley National Historic Park.
Stan F. Styles

Fort Langley, British Columbia

Canadian National Railways

As seen here in August 1958, Fort Langley station is on the Canadian National main line, 28 miles east of Vancouver. Stopped at the station is a CNR traveling museum train containing railroading exhibits. The station is decked with flags and newly painted in a light cream with green trimmings, to contrast with the black shingled roof.
Stan F. Styles

Garneau, Quebec **Canadian National Railways**

Garneau depot is in the Champlain district on the CNR Montreal-Rivière à Pierre line, 106.3 miles northwest of Montreal. The small, two-story, wood-frame station with shingled walls and a gable roof is very well maintained, as seen in this view of August 15, 1965.
Max Miller

Georgetown, Ontario **Canadian National Railways**

The Georgetown station is on the Canadian National Railway's main line Halton subdivision and was originally built by the Grand Trunk Railway of Canada. Originally incorporated by the Toronto and Guelph Railway in 1851, and amalgamated with the Grand Trunk Railway in 1853. The line was built in 1856, the same year the Georgetown station is believed to have been built. The station originally had a freight room extension which was removed in 1904 when extensive alterations were made. The grey cut stone building is still in active service as a commuter station serviced by GO Transit commuter service between Georgetown (now called Halton Hills) and Toronto that went into operation in April, 1974. This view shows the station as it looked on October 16, 1970.
George B. MacKay

Gananoque Junction, Ontario

Canadian National Railways

This station, seen in a view taken in the early 1940s, is on the Kingston subdivision, Rideau area, formerly the Gananoque subdivision. Erected in 1900, the station was built by the Thousand Island Railway, which was first incorporated as the Gananoque & Rideau Railway Company on February 15, 1871, but changed name on March 25, 1884. The spur line to Gananoque from the Grand Trunk Railway main-line junction opened for traffic on January 1, 1884. The spur was operated by the GTR from March 1, 1910, and the Thousand Island's stock was acquired by the GTR on March 6, 1911. The Grand Trunk amalgamated with the CNR on January 31, 1923.
Canadian National Railways

Gananoque Junction, Ontario

Canadian National Railways

This view of May 15, 1970 shows the station altered from the time of the previous photograph. That the roof bracketing has vanished and that the walls, previously timbered, are now shingled do not damage and may even improve the look of the station, but the same cannot be said for the boarding-up of tower windows and the disappearance of the beautiful gardens.
George B. MacKay

Grand Beach, Manitoba **Canadian National Railways**

The Grand Beach depot, built originally by the Winnipeg & Northern Railway Company in 1914, is located about 65 miles north of the city of Winnipeg. The wood-frame depot has a plank platform that is both wide and long. CNR No. 5089 was rolling into the station when this photograph was taken on July 11, 1949.
Stan F. Styles

Grand Beach, Manitoba **Canadian National Railways**
When, also on July 11, 1949, this view was made of a CNR train ready to leave on its 6 p.m. return trip, tourists regularly traveled to Grand Beach via train because the automobile roads were poor. Grand Beach is a popular resort on Lake Winnipeg with its fine white sand beaches. Today a highway to the once-small town has greatly increased its population and commercial atmosphere, and was probably the cause for the discontinuance of rail service in the 1960s.
Stan F. Styles

Grandview, Manitoba **Canadian National Railways**
Built in 1903 by the Canadian Northern Railway, this trim, little station is located 30 miles west of Dauphin. Exterior walls are painted white and the doors bright sky blue, while the roof is of dark blue shingles. The station was still in active service when this view was taken in August 1968.
Stan F. Styles

Inverness, Nova Scotia **Canadian National Railways**

The hip roof is rather awkwardly extended to form a trackside shelter on the Inverness depot. Outside the bay window at the passenger end is a plank platform, but there is none at the freight end. The depot was used for freight-only service by June 20, 1970, when this photograph was taken.
George B. MacKay

Kwinitsa, British Columbia **Canadian National Railways**

This station looks all the smaller for the coast mountain range in the background. The station's admirable bay penetrates the roof to continue in the dormer. July 1975 is when this view was made.
Stan F. Styles

Kwinitsa, British Columbia **Canadian National Railways**

Built in 1913 by the Grand Trunk Pacific Railway, this station is on the Prince George-Prince Rupert line. In this view, also dating from July 1975, CN No. 9179 is pulling into the station.
Stan F. Styles

Landis, Saskatchewan **Canadian National Railways**

The Grand Trunk Pacific built the Landis depot in 1911. The stucco structure has a half-octagonal bay that is contained in the dormer and an extended, unbracketed roof whose eaves act as a shelter. The depot, seen in this August 1968 view, remains in service.
Stan F. Styles

Lashburn, Saskatchewan **Canadian National Railways**

The Lashburn depot is on the North Battleford-Edmonton main line, 64.3 miles west of North Battleford. This two-story, stucco depot is of standard design and typical of many others still standing in the province. The late evening view looking down the line was taken in August 1972; the depot was then still in service.
Stan F. Styles

Makinak, Manitoba **Canadian National Railways**

This station is on the McCleary-Firdale branch line in southwest Manitoba, 156 miles west of Winnipeg. The two-story, stucco building with living quarters on the second floor is another variation on a standard design that can still be frequently spotted in the Canadian western prairie region. The station, with its wood-plank platform, is well maintained as seen in this August 1968 view.
Stan F. Styles

Malton, Ontario **Canadian National Railways**

The line began as the Toronto & Guelph Railway, incorporated in 1851. This company and others were in 1853 amalgamated into the Grand Trunk Railway Company. When the line reached Malton in 1854, the first Malton station was built. The line opened in 1856. The station was remodeled in 1903, but in 1914 the GTR, at the cost of $2752, built the station seen here. Canadian National Railways later acquired the property. The combination station, of brick over a wood frame, had come to be used mainly for freight by June 1973, when it was demolished as part of the preparations for the new GO Train service. The building is pictured as it looked on November 5, 1972.
Julian Cavalier

Malton, Ontario **Canadian National Railways**

The shelter, a type of structure that has been used in railroading since the early days, is here updated in the new Malton passenger shelters, photographed on March 24, 1974, before the GO Train commuter service began on April 29, 1974. Trains depart from Georgetown and head south, stopping along the line at several stations including Malton and finally arriving at Union Station, Toronto, before heading back to Georgetown. The original, Malton station site is about a quarter-mile north, close to where the freight car is seen in the distance. Modern shelters like those pictured may well be the future replacements of many existing stations in North America.
Julian Cavalier

Marshall, Saskatchewan **Canadian National Railways**

Marshall depot is on the North Battleford-Edmonton line, 72.6 miles west of North Battleford and 181.4 miles east of Edmonton. The stucco station is of a standard design and very similar to the Lashburn depot, which is the next depot east on the same line. This view looking down the tracks was taken in August 1972.
Stan F. Styles

Marshall, Saskatchewan **Canadian National Railways**

This sunset view of the Marshall depot looks in the opposite direction than does the accompanying daytime view. The depot has living quarters on the second story, and was still in service when this view was taken in August 1972.
Stan F. Styles

nastery, Nova Scotia **Canadian National Railways**

s depot is 123.8 miles from the city of Sydney, Nova tia. The wood-frame building contains passenger and ght facilities, as well as living quarters on the second floor. extra height of the freight room floor and the raised tform at left allow, *with an interceding baggage cart,* for ct loading of railroad cars. Trains stop on signal at this ion, still in active service when this view was taken on e 20, 1970.
rge B. MacKay

Montmagny, Quebec **Canadian National Railways**

Montmagny station is on the south side of the St. Lawrence River just east of Quebec City. A relatively plain first floor in brick supports a thick gambrel roof with shingled gables and elaborately decorated dormers. This photograph was taken on June 10, 1970.
George B. MacKay

Morris, Manitoba **Canadian National Railways**

On the Winnipeg-Minneapolis-St. Paul north-south route, Morris station is 40 miles south of Winnipeg. Similar in design to, but smaller than, the CNR Edson depot, Morris station is used by the Canadian National, Great Northern, and Northern Pacific Railroads. The two-story, stucco station at Morris was still in service when this August 1972 view was made.
Stan F. Styles

Neepawa, Manitoba **Canadian National Railways**

This building of painted brick originally had living quarters on the second story. The damaged corner of the eaves seen in this August 1968 view was the result of an accident, not of deterioration. The station was in active service when this view was taken.
Stan F. Styles

North Sydney, Nova Scotia **Canadian National Railways**

The North Sydney depot is on the east tip of Cape Breton Island, 15 miles northwest of Sydney. Ferryboat service is available between North Sydney and Port-aux-Basques. The walls are brick to one-third of the way up the windows, above which is stucco. A wood-plank platform is at trackside. The building, still well maintained and in active service, is seen in a view taken June 20, 1970.
George B. MacKay

Parkdale, Toronto, Ontario **Canadian National Railways**

The Parkdale station was built in 1878 by the Grand Trunk Railway to replace a smaller, ornate station building that existed just south of it and also called Parkdale. In the Fall of 1976 the Parkdale station was scheduled for demolition until a Save Our Station Committee was formed to press for its relocation and preservation. This proved successful and the old station was moved to a nearby location at a cost of several thousand dollars the Committee had raised. Plans to restore and preserve it as a historic site were ended on October 17, 1977 when the station was gutted by fire believed to have been started by transients. This view shows the Parkdale station when it was at its original location and still in active service as seen on May 29, 1971.
Julian Cavalier

Penobsquis, New Brunswick **Canadian National Railways**

Trains stop at signal at Penobsquis station, which is 38.1 miles from Moncton, New Brunswick. The *charming bravura* of both the window trim and the roof braces enhance this board-and-batten structure, seen here in a photograph made June 18, 1970.
George B. MacKay

Pictou, Nova Scotia **Canadian National Railways**

When this view was taken in about 1914, Pictou station was still part of the Intercolonial Railway, which built it. The two-story, brick building was used mainly for passenger service, but had a small baggage room at one end.
Cavalier Collection

Pictou, Nova Scotia **Canadian National Railways**

Pictou station was acquired by the CNR in the early 1920s. High windows are set within the elaborate gables and dormers. Though still sound, the building, seen in this view of June 19, 1970, no longer handles passenger service.
George B. MacKay

Portage-la-Prairie, Manitoba **Canadian National Railways**

This station was built in 1907 by the Grand Trunk Pacific Railway, now part of the Canadian National Railways. This view shows CN No. 5135 and CN No. 5090 leaving Portage-la-Prairie on a sunny day, August 31, 1950.
Stan F. Styles

Port Elgin, New Brunswick **Canadian National Railways**

Port Elgin depot is 58 miles east of the city of Moncton. The brick station has a stucco finish down to the line of the window sills. With its extensive, wood-plank platform, it was originally used for passenger traffic, as seen in this view of June 18, 1970.
George B. MacKay

River John, Nova Scotia **Canadian National Railways**

The River John station looks almost like a residence, and did indeed have living quarters on the second story. Coated with straight and zigzag shingles, the building gains verticality through its high dormer windows and tall chimney. Several other stations in Nova Scotia are of similar design.
George B. MacKay

Rothesay, New Brunswick **Canadian National Railways**

Located 80.5 miles from Moncton, this station was built in 1858 by the European & North American Railway. At upper right in this photograph taken on June 23, 1970 can be seen the sunporch that had been added after the stations construction.
Stan F. Styles

Sarnia, Ontario **Canadian National Railways**

A commuter train has stopped at Sarnia station in July 1974. Currently in active service and well-kept, the station is of painted brick, with white wood trim, braces, windows, doors, and gutters.
John Uckley

Sarnia, Ontario **Canadian National Railways**

This view was also taken in July 1974. At the south entrance to Lake Huron on the Toronto-Chicago main line, the station is 178.8 miles west of Toronto.
John Uckley

Shubenacadie, Nova Scotia **Canadian National Railways**

In this view of Shubenacadie station taken in the 1880s, the train and station crews (numbered) together with unidentified locals pose for the photographer. On the gingerbread second story of this wood-frame building are living quarters for the agent and his family. Note the large lantern that hangs between the windows of the first-floor bay and lit the platform area.
Canadian National Railways

GREENVILLE, CALIFORNIA Western Pacific

This is a view of the rustic-styled frame station at Greenville, California, as it looked November 10, 1975. On September 8, 1975 a notice to the public was placed in view on the station that, effective that date, the Western Pacific Railroad Company had reduced its agency station at Greenville, Plumas County, to nonagency status. The building is on a general rectangular design plan with a central two-story section and shorter single-story wings on each end on the longer central axis. The exterior walls are mainly shingled siding with a redwood base up to the lower window sills and a light yellowish color on the wood on the gables. The station is located on the "Inside Gateway" line north of Keddie, California.

Photo by Henry E. Bender, Jr.

McADAM, NEW BRUNSWICK, CANADA Canadian Pacific
A late evening sun casts its soft glow on one of Canada's last remaining combination station-hotels, as seen in this view of September 19, 1973. The building was built in 1900, with the last major changes made to it in 1910 by the addition of four gables and a pavilion in a style similar to the original design. This building was originally built by the European & North American Railway Company, which was taken over by the St. John & Maine Railway Company, which in turn was leased to the Canadian Pacific for a period of 999 years. It remained in use until the late 1950s. The guest rooms are still intact, and part of the former hotel facility now serves as a bunkhouse for operating employees. McAdam is located at the intersection of the main line extending between Montreal and St. John and branch lines extending northward to Woodstock and Edmonton and southward to St. Andrews and St. Stephen. St. Andrews has been a popular vacation resort for many decades, and the McAdam station-hotel afforded travelers the opportunity to stay over for brief periods when changing trains. The hotel was closed when public patronage on the trains declined. The building has a chateau influence in its design and is one of the finest buildings of its kind still existing in Canada.

Photo by Wayne Schenk

GRANDVIEW, MANITOBA, CANADA Canadian National
The station at Grandview is shown as it looked in August 1968 with the color combinations in the painting scheme as used by Canadian National. Note the blue colors used for the freight doors and the CN red used for the passenger or office entry door. Windows have white for the pane framing. Black is used to border the windows as well as for the doors and roof overhang braces—all set against the lighter exterior wall color that contrasts with the dark blue of the roofing.
Photo by Stan F. Styles

GLENWOOD SPRINGS, COLORADO Denver & Rio Grande Western
This is a picturesque view of the Glenwood Springs station as it looked on a winter day, December 29, 1974. Basically of cut grey stone construction, the station features a pair of towers flanking a main entry on the street side built of brick contrasting with the stone of the main building design. The station was still in active service when this view was taken.
Photo by Henry E. Bender, Jr.

BIRMINGHAM, MICHIGAN Grand Trunk Western
The Birmingham station was built in 1931 and is still in active service today, as seen in this view of February 1975. The station is basically of brick construction with a concrete facade entry on the street side and timbers in the design of the second-story gables that have a patterned brick design on the wall facings. The modern roofing is multicolored but was originally of slate. Across the tracks from the main two-story building is an enclosed platform shelter generally of a similar design to the main building in its roof style and brick construction. An underground tunnel leads from the main building to the platform shelter.

Photo by John Uckley

Spence's Bridge, British Columbia
Canadian National Railways

Spence's Bridge depot is on the Canadian National main line, northeast of Vancouver. The stucco depot is a variation on a standard design and similar to the Fort Langley depot. In its setting of rugged mountains, Spence's Bridge depot was photographed in April 1963.
Stan F. Styles

St-Bruno, Quebec **Canadian National Railways**

This quaint, little depot, seen on September 11, 1971, looks as if it belongs under a Christmas tree. Similar towers with small upper windows and sharply painted roofs are to be found in the Welland Junction and Georgetown stations in Ontario. Stations similar in overall design are at Danville and Pointe Claire in Quebec.
George B. MacKay

Ste-Anne-de-Beaupré, Quebec Canadian National Railways

Located 20.6 miles from Quebec City, the massive, cut gray stone station is worthy of its site in front of the great church. Even when wood is brought into play in the exterior staircase at the side, the design is still powerful. This view was made on October 11, 1971.
George B. MacKay

St-Léonard, New Brunswick Canadian National Railways

This station, in the north central part of the province, is 205.9 miles from the city of Moncton. The station is of brick, with the upper half of the walls in stucco. A long rectangle with hip roof is divided into two wings by a higher central block with gable roof. The station is kept neat and trim, as evidences this view of June 24, 1970.
George B. MacKay

Ste-Anne-de-Beaupré, Quebec Canadian National Railways

This building *is the essence* of late French Gothic architecture with rough stone finish contrasting with smooth, window surrounds, elegant doorway arch, and plain hip roof. Such a design easily overcomes the modern additions of an overhead garage door and a television antenna.
George B. MacKay

St-Philippe-de-Néri, Quebec **Canadian National Railways**

Similar in design to the Montmagny station, the shape of this station is narrower and higher. Both stations have massive gambrel roofs, but on this station the dormer roofs extend further beyond the window line. This photograph was made on June 11, 1970.
George B. MacKay

Stratford, Ontario **Canadian National Railways**

On the CNR Toronto-Windsor main line, Stratford station is 86.6 miles west of Toronto. The brick station rests on a base of cut stone that rises to the first-floor window sills. The lower story was intended for passenger facilities, and the upper for offices, while a single-story freight building was put at one end of the station. Still in service, the station is seen in a view of July 21, 1958.
Howard W. Ameling

Stratford, Ontario **Canadian National Railways**

There has been a station at Stratford for well over a hundred years. Stratford had one of the stations at which the young Thomas A. Edison worked as a railroad dispatcher. The station shown here, where No. 5079 is seen pausing in April 1958, was built by the Grand Trunk Railway and acquired in the 1920s by the CNR. The battlemented tower that makes the station seem so right for a city bearing the name of Shakespeare's birthplace has been removed.
Howard W. Ameling

St. Williams, Ontario **Canadian National Railways**

This splendid old photograph shows the St. Williams station in 1909. The upper floor probably held living quarters for the agent, who may have had to draw his water from the outdoor pump at the left of the picture. The wood-frame station was built in 1887 by the South Norfolk Railway Company, which had been incorporated on June 23 of that year. Grand Trunk Railway acquired by sale, the stock of the company, on June 11, 1888. The line was opened from Simcoe to St. Williams and Port Rowan on June 30, 1889. Grand Trunk amalgamated with Canadian National Railways on January 31, 1923.
Canadian National Railways

Sussex, New Brunswick **Canadian National Railways**

The Intercolonial Railway, which built it in 1913, still operated Sussex station when this view was made in 1916. CNR absorbed Intercolonial and this station in the early 1920s. The station contained both passenger and freight facilities. The freight house is seen at the extreme left.
Cavalier Collection

Sussex, New Brunswick **Canadian National Railways**

The passenger end of the station is in the foreground of this picture, taken on June 18, 1970. The building is of brick with a low base of stone.
George B. MacKay

Sussex, New Brunswick **Canadian National Railways**

This view, taken from the freight end of the building, was made in June 1970. Except for painting to the CNR colors and new roofing, the station, still in use, has changed very little from when it was built.
George B. MacKay

Tatamagouche, Nova Scotia **Canadian National Railways**

This view was made on June 19, 1970. Since the building was erected, some additions that increase the comfort and convenience of the second-floor living quarters have sprung up on the roof of the baggage room.
George B. MacKay

Truro, Nova Scotia **Canadian National Railways**

The Intercolonial Railway built this station, but in 1918, when this view was made, all of the government-controlled railways in the Maritime Provinces were under the name Canadian Government Railways, explaining the title in the upper left-hand corner of the picture. The station was acquired by the CNR in the early 1920s.
Cavalier Collection

Truro, Nova Scotia **Canadian National Railways**

As can be seen from this view of June 22, 1970, the station has changed little over the years. The large, brick building was equipped for both passenger and freight service; freight and baggage areas were each housed in one of the single-story wings flanking the two-story central section. An unusual feature is the second-floor bay window that is not repeated below. The square tower remains today with its original roof.
George B. MacKay

Wallace, Nova Scotia **Canadian National Railways**

Wallace station, seen here in a mid-June 1970 view, is a mirror copy of Tatamagouche station. Similar buildings in Nova Scotia are also at Oxford, Oxford Junction, and Pugwash.
George B. MacKay

Westville, Nova Scotia **Canadian National Railways**

The Westville depot was originally part of the railway of the Acadia Coal Company, but now belongs to the CNR. Seen in this June 18, 1970 view, the brick building with wood-plank platform is now used mainly for freight service.
George B. MacKay

Whitby, Ontario **Canadian National Railways**

Originally built by the Grand Trunk Railway in 1903, Whitby station was acquired in the early 1920s by the CNR. In this view made during winter in the late 1960s, the station is seen being moved from its original site to a new location a few blocks to the south.
Canadian National Railways

Whitby, Ontario **Canadian National Railways**

This view of October 16, 1970 shows Whitby station on its new site. The building has been expertly restored and now serves as the Whitby Art Gallery.
George B. MacKay

Bassano, Alberta **Canadian Pacific Railway**

Bassano depot is located on the Canadian Pacific main line east of Calgary. The building derives from one of the railroad's standard designs, and is similar to its station at Stavely, Alberta, the main difference being that the station pictured here is longer in the office section. Bassano station was still in service when this July 1969 view was taken.
Stan F. Styles

Biggar, Saskatchewan **Canadian Pacific Railway**

Biggar depot was still in service, but for freight only, when this photograph was made in August 1968. This depot also is a variation on a standard design, and as such is related to the Cut Knife depot in the same province. The freight room at Biggar, almost double the length of that at Cut Knife, achieves its charm through its very simplicity.
Stan F. Styles

Brandon, Manitoba **Canadian Pacific Railway**

Pulling 14 cars, Royal Hudson No. 2844 arrives at Brandon station. At the extreme right of the photograph, loaded baggage trucks await the train.
Stan F. Styles

Brandon, Manitoba **Canadian Pacific Railway**

Lead by No. 2844 westbound, the train receives and discharges passengers at Brandon station. Note the overhead walkway that was installed so people could safely cross the tracks. Like the previous picture, this one was taken on sunny September 1, 1950.
Stan F. Styles

Bristol, New Brunswick **Canadian Pacific Railway**

Now used for freight-only service, the Bristol depot is well maintained, as seen in this June 23, 1970 view. A concrete platform is at the front of the wood-frame building, whose roof at trackside projects somewhat awkwardly, as does the similar roof on the Canadian National depot at Inverness, Nova Scotia.
George B. MacKay

Colonsay, Saskatchewan **Canadian Pacific Railway**

Stations built in the standard design used at Colonsay are no longer as plentiful as they were once. The wood-frame, combination (passenger-and-freight) station displays an interesting variation in the use of double doors instead of the usual single sliding door. The eaves are particularly wide at this depot. The station was still in active service when this view was made in August 1968.
Stan F. Styles

Coquitlam, British Columbia **Canadian Pacific Railway**

Both arms out says stop to CPNo. 914, a D10 4-6-0 on a way freight from Vancouver. Seen in this September 1955 view, the wood-frame depot on the CPR main line, 16.5 miles east of Vancouver.
Stan F. Styles

Cut Knife, Saskatchewan **Canadian Pacific Railway**

This depot is another from one of the Canadian Pacific's wood-frame designs, and is similar to the depot at Biggar. Living quarters are on the second floor. This photograph was made on August 1972.
Stan F. Styles

Cypress River, Manitoba **Canadian Pacific Railway**

Originally a combination depot, Cypress River depot was used for only freight service when this view was taken in August 1968. The large, wood-frame building with living quarters above is based on a standard design, that can still be found at different sites in the western provinces.
Stan F. Styles

East Coulee, Alberta **Canadian Pacific Railway**

The East Coulee depot is in one of the CPR's standard designs. Built in 1914, the depot contains a waiting room and ticket office; the living quarters for an agent are on both floors. The depot, seen in this August 1968 view, is now used only for freight service.
Stan F. Styles

Exshaw, Alberta **Canadian Pacific Railway**

Set amid the Rocky Mountains, Exshaw depot is now used for freight-only service. Yet another variation on a Canadian Pacific standard design with living quarters on the second floor, the depot is in excellent condition as seen in this view of August 1964.
Stan F. Styles

Field, British Columbia **Canadian Pacific Railway**

At Field station in July 1952, No. 8410 prepares to leave for Vancouver. The station of wood and brick has a long, curved platform at trackside. The location is a very scenic area within the limits of Banff National Park.
Stan F. Styles

Field, British Columbia **Canadian Pacific Railway**

CP No. 5933 on the head end of No. 3 has arrived at Field station on April 22, 1951. Close to the eastern boundary of British Columbia, the station is on the CPR northern main line east-west route, 136.6 miles east of Calgary, Alberta and 505 miles west of Vancouver. The elevation is 5,044 feet above sea level.
Stan F. Styles

Gatineau, Quebec **Canadian Pacific Railway**

On the CPR's Montreal-Quebec-Ottawa line, Gatineau depot is 177.6 miles west of Montreal. This building, wood-frame with siding and shingles, is a combination type. The depot has a CPR express agency and was in active service when this view, taken from the office end, was made on May 2, 1970. George B. MacKay

Gatineau, Quebec **Canadian Pacific Railway**

The grade at the rear of this station is slightly below the level of the station and its long, paved platform. The roof at trackside bends up slightly to merge with the end canopies (each with a different pitch). This view, with the freight section in the foreground, was taken on October 24, 1970. George B. MacKay

Glacier, British Columbia **Canadian Pacific Railway**

This very handsome, log cabin of a railroad station is well suited to its Rocky Mountain setting. The CPR uses the building as a depot no longer, but as a bunkhouse now.
Stan F. Styles

Grand Falls, New Brunswick **Canadian Pacific Railway**

The narrow doors on the freight section of this combination station indicate it was expected to handle merely a light volume of freight. When this view was made on June 24, 1970, freight was in fact the only traffic the station still handled.
George B. MacKay

Ingersoll, Ontario **Canadian Pacific Railway**

The rugged-looking, little station features arched windows and cut stone. The station, now used for freight-only service, is still in good condition as seen in this view of March 23, 1975.
John Uckley

Ingersoll, Ontario **Canadian Pacific Railway**

Another view shows that, at the freight end, the windows and doors have been closed over. This photograph was also made in March 1975.
John Uckley

Lachute, Quebec **Canadian Pacific Railway**

This very handsome, Tudor structure is located 50 miles west of Montreal. A low, stone base (higher at the rear to allow for the grade) supports brick walls trimmed in uniformly cut stone. The roof on this combination station is lower at the freight end than at the passenger end. This view of April 11, 1970 shows the station still in active service.
George B. MacKay

Lake Louise, Alberta **Canadian Pacific Railway**

Located 116.6 miles west of Calgary, Lake Louise station is at 5044 feet above sea level. In accordance with its rustic setting, the station was designed as a log cabin. When this view was made in August 1964, the station was still in service for passenger service. Now, however, it is used instead for storage.
Stan F. Styles

Lethbridge, Alberta **Canadian Pacific Railway**

Lethbridge station in south central Alberta is on the Canadian Pacific's Medicine Hat-Lethbridge line. The first story of this combination station is of brick with a stone base. The wood, second story contains a rhythmic row of six identical dormers, intricately shaped, at trackside, and a tower at streetside. The station is still in active service as pictured in this view of June 1969.
Stan F. Styles

Lorette, Quebec **Canadian Pacific Railway**

Located 7.4 miles southwest of Quebec City, Lorette depot is on the Montreal-Quebec City main line. Following one of the railroad's standard designs, the 1914 depot has walls covered in sidings and shingles. When this view of April 29, 1967 was taken, the depot was still in active service.
Max Miller

Masson-Buckingham Junction, Quebec
Canadian Pacific Railway

On the Montreal-Ottawa line via the Montebello route, the combination depot is 105.6 miles west of Montreal. The roof, now in multicolored shingles, widely overshoots both ends of the long block, and an open gable shields the roof of the trackside bay. The depot is seen in active service in this view of October 24, 1970.
George B. MacKay

Medicine Hat, Alberta **Canadian Pacific Railway**

Built in 1906 and enlarged in 1911, the Medicine Hat station is on the Canadian Pacific's Medicine Hat-Lethbridge line 111.6 miles east of Lethbridge. At the first story, brick and pairs of tall, narrow windows surmount a low, stone base. Springing from the hip roof are a jovial assortment of towers and dormers. This marvelous station was still in active service when this view was made in July 1969.
Stan F. Styles

Midnapore, Alberta **Canadian Pacific Railway**

In 1910, this station was built at Midnapore, now part of the city of Calgary, as a train order office for moving trains over the line. Later it became an agency, but was closed in 1918. Mr. Cronk was the first and only station agent; he moved on to Cayley, Nanton, and Blairemore, and finally retired in 1951. Various caretakers occupied Midnapore station for a number of years, but it was finally boarded up. Mr. Paddy Bowman, R.F.P., Superintendent, Lethbridge, purchased it from the CPR in 1963 for $1. Today it is located a short distance from the entrance gate of the Heritage Park Society, Calgary. The total cost of moving and renovating the station and erecting a small ticket house was $24,000. The telegraph office was restored with equipment, furniture, and even a mannequin telegrapher. The upstairs was made into the park administration offices, which became so crowded that the historian was forced to share the telegraph office with the mannequin. However, in 1972, the administration office was moved to another building. At present, the offices upstairs are being used by Alberta Theatre Projects, a group that performs in the park's Canmore Opera House.

Ranson Photographers Ltd., Courtesy Heritage Park Society

Midway, British Columbia **Canadian Pacific Railway**

In the south central part of the province near the Washington state border, the wood-frame building follows one of the railroad's standard designs that has station facilities on the first floor and the agent's living quarters above. This scenic sunset view was taken in October 1974.
Stan F. Styles

New Westminster, British Columbia
Canadian Pacific Railway

This combination depot is of brick with stone base and trim, both materials having been painted. The tall, central block presents at trackside a pair of towers on either side of a section with sharply pointed gable. From the side spread lower wings, one each to accommodate baggage and freight. Windows are deep-set as seen in this October 27, 1971 view.
Harry Schimm; George B. MacKay Collection

3671
PORT MOOD

Papineauville, Quebec **Canadian Pacific Railway**

Located 84.8 miles west of Montreal, the station consists of a hip-roofed, two-story block with a one-floor, freight wing, whose roof continues across the rest of the station to form a trackside shelter. This well maintained station was still in active service when this view of October 25, 1970 was taken.
George B. MacKay

Saskatoon, Saskatchewan **Canadian Pacific Railway**

Although still a passenger stop, Saskatchewan station is mainly used today for freight service. Upon looking at this August 1968 view, the eye is drawn along the single-story freight wing to the massive, two-story main section with its bulky tower. A similar design, but without the extensive freight section, is found in the South Edmonton, Alberta station.
Stan F. Styles

Port Moody, British Columbia **Canadian Pacific Railway**

Semaphores up, indicating a clear board, CP No. 3671 on a transfer job between Coquitlam and Vancouver steams through Port Moody in this view of June 14, 1953. The wood-frame depot, now used for only freight, is of a standard Canadian Pacific design and still in good condition.
Stan F. Styles

Shepard, Alberta **Canadian Pacific Railway**

Built to one of the CPR's standard designs, this station stood originally at Shepard, an elevator point one mile east of Calgary. In 1970, the railroad sold the station for $5, after which it was moved to a site in the village across the tracks from the Wainwright Hotel, at the Heritage Park Society, Calgary. Of the cost of moving and restoring the station, $45,000 was paid by the Glenmore District Association, which upon dissolving in 1970 wanted to apply its remaining funds to a worthwhile community project in the Glenmore area. Park grants are covering the balance of the $56,000 total. Since the move to the park, the baggage room has been cut back, the space gained becoming a waiting room, a boat ticket office, and a rest room. The main area of the first floor was made into a snack bar and kitchen. Work on the upstairs rooms has not yet been finished.
John Patrick Photographers Ltd., Courtesy Heritage Park Society

Sicamous, British Columbia **Canadian Pacific Railway**

The Sicamous structure was a station-hotel designed by architect Edward Maxwell. It opened for service on May 1, 1900 in its very scenic setting amidst the mountains of British Columbia next to a body of water where one might envision it as a luxury liner in port. It was a magnificent structure in its time of frame and stucco construction with screened porches facing the water and a track platform on the opposite side where passengers could arrive and depart in a most convenient manner. The building no longer exists today being dismantled in the mid 1960's. This view shows the station as it looked in April, 1963.
Stan F. Styles

Sicamous, British Columbia **Canadian Pacific Railway**

Built on the foundations of the original Sicamous station-hotel, this small station facility is a far cry from the grand structure that once stood here. This view, taken in August, 1969 shows a small frame structure with bay window and flat roof. Note the gravel platform area where the original platform was once located.
Stan F. Styles

South Edmonton, Alberta **Canadian Pacific Railway**

Closely related in design to the Saskatoon station, South Edmonton is a two-story, brick building with cut stone base and a big, octagonal tower. Built as a passenger station, it was used for freight alone when this photograph was taken in August 1968.
Stan F. Styles

Stavely, Alberta **Canadian Pacific Railway**

Stavely depot is on the Calgary-Lethbridge line, 72 miles from Calgary. The combination station is in one of the Canadian Pacific's standard designs. The wood-frame building has living quarters on the second floor, and at front a plank platform that slopes at either end to allow easier access. The station was still in active service when this August 1964 view was taken.
Stan F. Styles

Ste-Agathe des Monts, Quebec **Canadian Pacific Railway**

On the CPR's line between Montreal and Mont-Laurier, Ste-Agathe depot is 69.4 miles north of Montreal. Trains approaching from one direction are easily spotted from within the graceful round bay with conical tower. This well-shaped station was still in active service when this April 25, 1970 view was taken.
George B. MacKay

Trenton, Ontario **Canadian Pacific Railway**

This rather stripped-down model of an Elizabethan station was used mainly for freight by June 16, 1972, when this photograph was made. The massive yet intricate piers for the canopies demonstrate that the architects of this station's period, who worked in the Elizabethan style, were influenced by the turn-of-the-century interest in Chinese and Japanese design.
Howard W. Ameling

Unity, Saskatchewan **Canadian Pacific Railway**

Unity depot is a standard combination station with living quarters above. Such wood-frame, shingle-covered stations were erected abundantly in the Canadian West, but are gradually being closed and demolished. This depot was in use for freight-only service when this view was taken in August 1968.
Stan F. Styles

Woodstock, New Brunswick **Canadian Pacific Railway**

A concrete base supports the upper walls with their alternately protruding and receding bands of brickwork. A wide porte-cochere is at one end, and a fine bay projects at trackside. The paved platform is on a level with the parking lot. The well maintained station had come to be used only for freight when this view of June 24, 1970 was taken.
George B. MacKay

CANAL WINCHESTER, OHIO Chesapeake & Ohio

The Canal Winchester depot is a unique late nineteenth-century design wood frame structure that has been restored and repainted and is now used by a real estate company. The octagonal portion in the foreground once was the waiting room, with an office and baggage-freight room behind this, as seen in this view on the street side. The windows and door are arched, and the roof overhang is graced by a carved and curved wood design set in series around the building. Note the small hand water pump near the platform steps. This view shows the depot as it looked June 10, 1973.

Photo by David P. Oroszi

GRASS LAKE, MICHIGAN Michigan Central

The Grass Lake depot was built in 1887, originally for the Michigan Central Railroad, and it is now occupied by a printing company, as seen in this view of June 16, 1975. The small stone depot has changed very little since it was built and features a small round tower, an arched bay window with gabled roof at trackside, an open porch at the rear, and several floor styles, all of which are not completely evident in this view. The fieldstone used in the building's construction does have some contrasting colors mixed with the grey natural stone. The former depot, built as a passenger station, still exists and was well maintained when this view was taken.

Photo by John Uckley

RED PASS JUNCTION, BRITISH COLUMBIA, CANADA Canadian National

The Red Pass Junction depot is located in the Rocky Mountains forty-four miles west of Jasper on the Canadian National main line, which forks northward and southward from this point. The depot is of a standard older design with stucco exterior walls and a two-story bay section at trackside. The red roofing is in contrast to the lighter exterior wall color, with green used on the window outer framing of one door, on the upper windows, on the bay windows, and on the screen door. Note also the CN red and blue color of the doors and the mixture of white on the trimmings of the freight door entry and the windows at the left, where the larger freight door may have once existed. The depot was still in use when this view was taken in September 1972.

Photo by Stan F. Styles

Ridgeway, Ontario, Canada Canadian National
Originally built by the Grand Trunk Railway, the Ridgeway depot is of late nineteenth-century design and of wood frame construction featuring a two-story tower extending up from the bay and an open roof covered waiting area, at the left in this view. Built for passenger service, this depot originally contained a waiting room at the right of the tower, an office and large ladies' room centrally located, and a baggage room and small fuel and storeroom to the left of the tower. This view shows the depot as it looked on a winter day, February 2, 1972.

Photo by Michael P. McIlwaine

EXSHAW, ALBERTA, CANADA Canadian Pacific
This is a picturesque view of Canadian Pacific's Exshaw station as it looked in August 1964. Note the picket fence for the small garden at the left and the red paint on some of the brickwork in contrast to the original buff color of the bricks. The station was still in active service when this photo was taken.

Photo by Stan F. Styles

Yahk, British Columbia **Canadian Pacific Railway**

Westbound CP No. 8709 is receiving orders "on the fly," as seen in this view of October 1974. The first story of this combination depot is of brick, while the upper half-story is shingled. Now used for freight-only service, the station is well-kept and in excellent condition.
Stan F. Styles

Yoho, British Columbia **Canadian Pacific Railway**

This diminutive depot is dwarfed even more by its setting in the Rocky Mountains, with Mt. Stephen seen in the background of this May 27, 1949 view. The tracks are actively above the floor level of the station, and the trackside area that serves as a platform must slant down slightly to reach the station.
Ernie Plant Photo; Collection of Stan F. Styles

NEWNAN

Cedartown, Georgia **Central of Georgia**

Cedartown depot, west of Atlanta, is on the Central of Georgia's Griffin-Chattanooga main line. The depot once housed division offices as well as passenger facilities, but currently is used for freight service only. This lackluster depot has a brick first floor and a stucco second story. The second-floor bay looks out of place without support beneath it, but large windows and the roof's diagonal texture work for the appearance of the station. This photograph was made on March 19, 1970.
William F. Rapp, Jr.

Somerville, New Jersey **Central Railroad of New Jersey**

Located in the countryside, 34 miles from Jersey City and 28 from Newark, Somerville station was built in the 1880s from the designs of architect Frank V. Bodine. Fashioned of light gray sandstone and still in possession of its original slate roof, the station shows the architecture of the period making the transition from the earlier rustic structures to the new monumental construction. Although some changes have been made to the station, it still looks almost as it did when new, and is well-kept and in good condition as seen in this view of October 1970.
Herbert H. Harwood, Jr.

Newnan, Georgia **Central of Georgia**

Located 39 miles west of Atlanta, the Newnan station sits at the junction of the main line of the Atlantic & West Point and Central of Georgia railroads, both of which use the station. From the corner bay, the L-shaped building spreads one of its wings along each of the tracks. The station was still in service when this photograph was taken on March 20, 1970.
William F. Rapp, Jr.

Howell, Michigan **Chesapeake & Ohio**

The Howell depot is on the C&O main line, which runs between Detroit, Lansing, Grand Rapids, and Chicago. The first railroad through this community was the Detroit & Howell Railroad. That went into bankruptcy, and in 1879 a new company, the Detroit, Lansing & Northern Railroad, gained control. The line became part of the Pere Marquette Company in 1900, and in 1947 the Chesapeake & Ohio took control of that company's lines in Michigan and Canada. This depot was completed on January 15, 1869, and though passenger service to this station ended in 1971, it is still used as a freight office by the C&O. This view shows the depot as it looked in June 1974.
John Uckley

Kingsville, Ontario, Canada **Chesapeake & Ohio**

Kingsville station located near Lake Erie about 30 miles from Windsor, Ontario. This was once the single track Pere Marquette R.R. line that ran between Windsor and St. Thomas, Ontario where passenger train service ended in the mid 1930's. A part time agent still occupies the station and the waiting room is now a store room and kitchenette for the maintenance of way crews. The stone structure is used for freight service only by the C&O and is still in excellent condition as seen in this view of 1975.
John Uckley

Newport, Kentucky **Chesapeake & Ohio**

Located on the C&O main line, the Newport, Kentucky depot is 5 miles southeast of Cincinnati, Ohio. Having a plan in the shape of a T, the station is particularly interesting for its delicate use of wood, as in the board-and-batten exterior wainscoting that supports rectangles filled by diagonal boards, in the roof framing and brackets, and in the fan beneath the trackside gable. The station was still in service when this view was taken in June 1974.
David P. Oroszi

Peru, Indiana **Chesapeake & Ohio**

A variety of shapes—an octagonal bay at one end, a circular bay near the middle, a rectangular section at the other end—are present in this station, and the roof must be sufficiently elastic to cover all this. Now used for freight-only service, the brick-above-stone station is well maintained, as seen in this view of June 1974.
David P. Oroszi

Williamsburg, Virginia **Chesapeake & Ohio**

Georgian is the style of historic Williamsburg's most impressive buildings, and so it's appropriate for this station on the Newport News-Richmond main line, 48 miles southeast of Richmond. Symmetry is a mark of the style, and the station was originally built accordingly, but the extra sections at left in this streetside view were added later without detriment to the whole building. In excellent condition and still in active service, the station is seen in a photograph from July 19, 1970.
R. D. Patton

Bellevue, Nebraska **Chicago, Burlington & Quincy**

Believed the oldest railroad depot in Nebraska, Bellevue was built in 1869-1870 by the Omaha & Southern Railroad, that later became part of the Chicago, Burlington & Quincy, which, in turn, became the Burlington Northern. This view of March 6, 1967 shows the depot on its original site before a move to Haworth Park left the one-story, wood-frame structure standing nevertheless next to the railroad tracks. The building is now a museum.
William F. Rapp, Jr.

Highland, Illinois **Chicago, Burlington & Quincy**

Suburban commuter stations often have, as does this one on the line between Cook and DuPage counties 16.4 miles west of Chicago, a larger building, with waiting room, ticket office, and other facilities, on one side—usually the inbound one—of the tracks, and a smaller building, often no more than an open shelter as is this, on the other—usually the outbound side. The larger building, then, does all the ticket selling and is the waiting place for people going into the city. The smaller structure is the waiting area for the relatively far fewer who want to depart the station for points further down the line and for those arriving home who want to be picked up in a private vehicle.
Burlington Northern

Highland, Illinois **Chicago, Burlington & Quir**

This stone building with the tin roof was built in mid-1870s and refurbished in 1974. A lower story, readily evident in the other picture, allows for a fall in terrain. After the merger that created the Burlington Nor ern, the exterior trim was repainted in the railroad's li green, a change from the earlier dark green of the Chica Burlington & Quincy.
Burlington Northern

Hot Springs, South Dakota **Chicago, Burlington & Quir**

For many years the waters of this small community in Black Hills have attracted tourists. This solid little station gray cut stone sports a square tower. Built by the C.B.& the station was later acquired by the Burlington Northe and in its later days, as an active depot, was used wholly freight. Today it houses the Chamber of Commerce, as it on July 30, 1963 when this view was made.
William F. Rapp, Jr.

Prairie du Chien, Wisconsin **Chicago, Burlington & Quincy**

On the Burlington Line's Chicago-Minneapolis main line 188 miles southeast of St. Paul, Minnesota, the depot at Prairie du Chien is still in active service as seen in this July 6, 1967 view. Originally built by the Chicago, Burlington & Quincy Railroad, the brick station is in the rather bleak style favored by the mid-20th century. (*See also* the Chicago, Milwaukee, St. Paul & Pacific station in the same community.)
William F. Rapp, Jr.

Red Oak, Iowa **Chicago, Burlington & Quincy**

Set in the western part of the state, Red Oak station is on what is now the Burlington Northern's Chicago-Omaha main line 50 miles west of Creston, Iowa. Built by the Chicago, Burlington & Quincy, the brick building was, in the days of steam, not only a busy passenger station but also an important train order office. The station was still in use when this view was taken on December 5, 1967.
William F. Rapp, Jr.

Wymore, Nebraska **Chicago, Burlington & Quincy**

This structure makes a point of its material being wood by using long, straight sticks to cover vertical corners, to conspicuously support the roof gables, as brackets for the exterior shelter around the first story, and in frames for the tall windows. Built by the C.B.&Q. to house a large passenger waiting room, a ticket office, and the Wymore Division offices, the depot is seen here in a view of March 21, 1968.
William F. Rapp, Jr.

Browntown, Wisconsin
Chicago, Milwaukee, St. Paul & Pacific

This primitive depot is a two-story, board-and-batten structure. The branch-line station originally housed a passenger waiting room, office, and freight room. When this view was taken on July 7, 1967, the building also contained a Railway Express Agency.
William F. Rapp, Jr.

Elk Point, South Dakota

Chicago, Milwaukee, St. Paul & Pacific

Located 21 miles north of Sioux City, Iowa, Elk Point depot is in the southeast corner of South Dakota, close to the Missouri River. The little, wood-frame structure is believed to have been the first depot the Chicago, Milwaukee, St. Paul & Pacific Railroad built here. The depot was used for freight-only service when this view was made on December 7, 1967. William F. Rapp, Jr.

Kittitas, Washington

Chicago, Milwaukee, St. Paul & Pacific

Made in September 1975, this is a twilight view of Kittitas depot on the Seattle-Spokane main line, 167 miles west of Spokane. The wood-frame, single-story station is now used only for freight.
Stan F. Styles

Missoula, Montana **Chicago, Milwaukee, St. Paul & Paci**

This station of imposing size, but perhaps insuffici ornament, is in western Montana on the Milwaukee Rc main line. The two-story building used originally for pass gers has good fenestration, especially on the second flc The features of the smaller tower make it a conden version of the taller. A smaller freight station is next to station, but by now both buildings, still well maintain handle only freight.
William F. Rapp, Jr.

Prairie du Chien, Wisconsin
Chicago, Milwaukee, St. Paul & Pac

At one end of the railroad's Madison-Prairie du Chien li this depot is 97 miles west of Madison. The building ha rough stucco finish and is decorated with bands of brick. T photograph was made on July 6, 1967.(*See also* the Chica Burlington & Quincy station in the same community.)
William F. Rapp, Jr.

Beaver Crossing, Nebraska **Chicago & North Western**

The dark band that surrounds the station about two-thirds up the walls and is also found at corners and around windows and doors, shows this station is not as unsophisticated as might be thought. The wood-frame depot, seen in a 1965 photograph, had living quarters for the agent, a freight room, a passenger waiting room, and a large office. It is on the Superior branch, which is now abandoned.
William F. Rapp, Jr.

Fennimore, Wisconsin **Chicago & North Western**

Fennimore depot is 76 miles west of Madison and is the last depot on the Madison-Fennimore branch line. At one time a combination depot, Fennimore is no longer used by the railroad. As seen in this July 6, 1967 view, the fine building is badly in need of repairs.
William F. Rapp, Jr.

Heron Lake, Minnesota **Chicago & North Western**

This depot was originally built by the Chicago, Milwaukee, St. Paul & Omaha Railroad; for many years, "the Omaha" was operated as a separate company. The wood-frame building is 150 miles south of Minneapolis and now used only for freight. The depot was still in active service when this view was made on November 24, 1968.
William F. Rapp, Jr.

Mountain Lake, Minnesota **Chicago & North Western**

This typical, small-town combination (passenger-and-freight) depot now handles freight alone. The depot, located 127 miles southwest of Minneapolis, has a new partial roof above the office area, as seen in this view of November 24, 1968.
William F. Rapp, Jr.

Sleepy Eye, Minnesota **Chicago & North Western**

Said to have been built to impress the citizens of the growing city in southern Minnesota, Sleepy Eye station is on the railroad's east-west Sleepy Eye-Marshall branch line, 58 miles west of Minneapolis. The handsome station of brick has contrasting white concrete window sills. Seen in a November 23, 1968 view, the station has a Railway Express Agency and is still in service.
William F. Rapp, Jr.

Winnetka, Illinois **Chicago & North Western**

No longer used for rail service, this striking building is today occupied by Zengeler Cleaners, one more example of the practically unlimited range of applications to which former railroad stations can be put. The Georgian requirement of symmetry was lived up to by dividing the facade into smaller symmetrical parts: the hip-roofed passenger section, the flat-roofed freight section, and its windowless continuation. On the Chicago-Milwaukee-Green Bay line, the station is 16.6 miles north of Chicago.
Henry E. Bender, Jr.

Fairbury, Nebraska **Chicago, Rock Island & Pacific**

This two-and-a-half-story, Italian station is of red brick with a tile roof. Built in the early 1920s to handle passenger traffic and house the offices of the railroad's Nebraska division, the station was in active service when this March 28, 1962 view was taken.
William F. Rapp, Jr.

Luverne, Minnesota **Chicago, Rock Island & Pacific**

The two-story, wood-frame station at Luverne is seen in this view of July 18, 1967. The second floor contains living quarters for the agent and his family. Ornate roof brackets, wooden window awnings, and a tiny balcony all add to the charm of this elegant station.
William F. Rapp, Jr.

Petoskey, Michigan **Chicago & West Michigan**

The Petoskey, Michigan station was built in 1892 by the Chicago & West Michigan Railroad who had purchased spacious grounds on Little Traverse Bay the year before for the erection of a new station building. This view shows the Petoskey station as it existed July 17, 1974. This trackside view faces the Bay.
Steve Salamon, David Oroszi Collection

Petoskey, Michigan **Chicago & West Michigan**

A streetside view of the Petoskey station showing the Bay in the background. This view of July, 1974 shows the building as it exists and leased to the Little Traverse Regional Historical Society, who have restored the building considerably since their occupation of it.
Steve Salamon, David Oroszi Collection

Whitehall, New York **Delaware & Hudson**

Near the New York-Vermont state line, Whitehall depot is 83.7 miles from Albany, New York. The station building is well behind and above the tracks, and is connected to them by a sloping, covered ramp that leads down to a trackside canopy. This view of June 27, 1969 shows the station and its ornate platform pole lamps in deteriorating condition.
Howard W. Ameling

Chama, New Mexico **Denver & Rio Grande Western**

The wood-frame, combination depot is viewed on June 5, 1960 when still operated by the D&RGW. Now owned by the Railroad Authority of Colorado and New Mexico, the station serves as operations headquarters of Scenic Railways, Inc., which runs the Cumbers & Toltec Scenic Railroad.
Henry E. Bender, Jr.

Durango, Colorado **Denver & Rio Grande Western**

The Durango depot was built in 1882 about the time the Denver & Rio Grande rail lines reached the town. It served as a division office on the D&RG for many years and later for both the D&RGW and RGS, the latter moving to another location in 1934. It is one of the largest stations serving narrow gauge trains and still exists in excellent condition. This view shows the station as it looked in August, 1961 and looking much as it did when originally built.
Stan F. Styles

La Jara, Colorado **Denver & Rio Grande Western**

Seen on May 26, 1962 from the baggage-room end, La Jara station has a design straight from the Old West. Among the features are the second-story, living quarters which open onto an enclosed terrace.
Henry E. Bender, Jr.

La Jara, Colorado **Denver & Rio Grande Western**

Stucco walls rise from a base of siding. In this November 13, 1949 view from the south end of the station is seen part of the Mikado D&RGW No. 473 pulling the famed "San Juan," which had its last run on January 31, 1951.
Ernest S. Peyton

Montrose, Colorado **Denver & Rio Grande Western**

Now used mainly for freight, Montrose station is 72.8 miles from Grand Junction. The Mediterranean station has a covered arcade at trackside. A standard gauge Mountaineer D&RGW No. 784 comes into Montrose station on a three-rail system that allows for standard gauge. Last run for the train was on October 24, 1959. This view shows how it was on March 30, 1950.
Ernest S. Peyton

Richfield, Utah **Denver & Rio Grande Western**

Although designed with a passenger waiting room and ticket office, the station today is used for freight-only service. Rows of wooden darts point downward from the eaves, and a freight platform with open shelter is seen at the extreme right of this view taken August 26, 1966.
Van W. Best; W. F. Rapp Collection

Avon, New York **Erie Lackawanna**

This wood-frame depot has an elaborately decorated track-side pediment. The depot is on the Attica branch, 34.8 miles east of Attica. When this photograph was taken on February 5, 1970, the depot was used only for freight.
Elmer Herbst Photo, Howard W. Ameling Collection

Basking Ridge, New Jersey **Erie Lackawanna**

Earlier a part of the Delaware, Lackawanna & Western, which became the Erie Lackawanna, this small commuter depot is on the Passaic and Delaware branch line. Of basic design, the building has cement exterior walls and a tile hip roof. The depot is well maintained and in active service when this view was taken on March 9, 1973.
William F. Rapp, Jr.

Chatham, New Jersey **Erie Lackawanna**

Referred to as the eastbound depot and containing the ticket office and baggage room, this station is seen in a March 9, 1973 view. Built in 1914, it was once on the Delaware, Lackawanna & Western, which is now the Erie Lackawanna. The brickwork, in alternating colors, is particularly interesting.
William F. Rapp, Jr.

Closter, New Jersey **Erie Lackawanna**

The small depot at Closter is a fine example of an old depot building that has been refurbished and is now being reused by a nonrailroad commercial business, in this case the Junction Gallery/Studio. As seen in this July 1974 view, the tower and other elements in a Victorian style give the building an interest that a new structure would be hard put to equal.
Herbert H. Harwood, Jr.

Little Falls, New Jersey **Erie Lackawanna**

On the Erie Lackawanna's Greenwood Lake division, the Little Falls depot is part of the Greenwood Lake-Boonton line and 18.4 miles west of Hoboken. The small, brick depot is used mainly for suburban passenger service. Built in 1900, the depot is seen in a view of March 10, 1973.
William F. Rapp, Jr.

Oradell, New Jersey **Erie Lackawanna**

Wrapped in a coating of shingles of varied shapes, this interesting depot was built in 1890 by the New Jersey & New York Railroad, a subsidiary then of the Erie Railroad. Today the Erie Lackawanna stops there for limited commuter service, but the building, now a gift shop, has no agent or waiting room. This photograph was made in August 1974.
Herbert H. Harwood, Jr.

Union Point, Georgia **Georgia Railroad**

This nondescript, wood-frame station has attached to it an extensive open shelter to accommodate long passenger trains. On the railroad's Augusta-Atlanta main line, Union Point depot is 76 miles west of Augusta. Built prior to World War I, the station is pictured in March 1966.
William F. Rapp, Jr.

Birmingham, Michigan **Grand Trunk Western**

The Birmingham station is located 17.7 miles west of Detroit on the Grand Trunk Western's Detroit-Pontiac-Durand-Chicago line. This newer station was built in 1931 with the first day of opening marking the opening of service of the railroad's commuter service between the Bush Street Detroit station to Royal Oak, Birmingham, and Pontiac in the summer of that year. This view of August 14, 1974 shows the street side entry.
John Uckley

Birmingham, Michigan **Grand Trunk Western**

A trackside view of the Birmingham station showing elevated passenger platform which extends along the tracks flanking the station building. The station is still in active service today as a commuter train passenger station facility. The GTW considers its Birmingham station as one of their finest passenger stations. A tunnel runs under the tracks to a long, narrow platform shelter for westbound trains.
John Uckley

Mount Clemens, Michigan **Grand Trunk Western**

By Independence Day 1859, the Detroit and Port Huron division of the Grand Trunk had been completed to Mount Clemens. This fine brick station was erected around the same time. It was at this station that, in 1862, the 15-year-old Thomas Edison saved 3-year-old Jimmy Mackenzie from being struck by a boxcar. Although Edison claimed to have had much time and little risk, Jimmy's father, J. U. Mackenzie, the Mount Clemens stationmaster, offered out of gratitude to teach the teenager to be a telegrapher, a profession that served him well during his early career. The station, altered a little over the years, is seen in a January 26, 1975 view.
John Uckley

)urand, Michigan **Grand Trunk Western**

ṡpier and Rohns were the architects for this station, which vas completed on October 17, 1905, only six months after ts predecessor had been destroyed by fire in April of the ame year. Located 67 miles northwest of Detroit, the station its at the intersection of two lines. Although the station, at irst glance, might seem oriented primarily toward the track unning parallel to its facade with projecting shelter and twin owers, it should be noted that no doors are seen there, and hat access is easier to the track at the station's side. The tation has served the Grand Trunk Western as well as the 'oledo, Ann Arbor & Northern Michigan Railway. Seen in a Aay 1974 view, the station is still in service.
ohn Uckley

Oxford, Maine **Grand Trunk Western**

The old Oxford depot, which no longer stands, was 41 miles from Portland. Note the lamps, probably kerosene, that hang on the walls; each lamp contains a reflector and is enclosed in a glass box to ward off the wind.
Cavalier Collection

Port Huron, Michigan **Grand Trunk Western**

The large, two-story station features fine brickwork, as witness the arches around the first-floor windows, the decoration across the top of the second floor, and the chimneys. On the GTW railroad's main line, 80 miles east of Durand, Port Huron station was an important point for passengers entering or leaving Canada. The building contained several offices and was headquarters for the Electric division to Canada, as well as holding extensive passenger facilities. This photograph was taken on June 28, 1972.
William F. Rapp, Jr.

:ortes, Washington **Great Northern**

· the Strait of Georgia, Anacortes depot is about midway ·een Seattle and the Canadian border. In a style that ıt be described as Elizabethan gone modern, the building vertical cement panels upon a brick base. Taken from t was originally the freight end, this view shows the depot looked on September 8, 1968. Today the depot is used ısively for freight.
ry E. Bender, Jr.

Bainville, Montana **Great Northern**

A July, 1971 view of the Bainville station showing its more modern 20th Century lines of architecture. The station was remodeled in 1957 and is basically of wood frame construction with a large roof covering the entire structure. The interior contains an office, waiting room and large freight room, as well as several other smaller rooms. The cedar lap siding of the exterior walls is typical of many other stations built during this period.
Stan F. Styles

h Lyon, Michigan **Grand Trunk Western**

wood-frame building sports a conical roof, popularly d a "witch's hat." Located about 35 miles northeast of oit, the station stands within one of the angles made by :rossing of Grand Trunk and Chesapeake & Ohio tracks. se by both railroads, the depot now handles only freight. photographer's father, Albert J. Uckley, takes a coffee k in this view of October 1975.
. Uckley

Eugene, Oregon **Great Northern**

Located 142.8 miles south of Portland, this is the southern terminal of the Eugene line of the Oregon Electric Railway, a subsidiary of the Great Northern and Northern Pacific railroads. The brick, classical building is now occupied by a branch of the Oregon Museum of Science and Industry, as seen in this September 12, 1968 view.
Henry E. Bender, Jr.

Tolna, North Dakota **Great Northern**

This depot is in a design that is standard for Great Northern depots. It is similar to the railroad's depot at Warwick, North Dakota. Tolna depot, on the Nolan-Devils Lake line, 66 miles north of Nolan, was used for freight-only service when photographed on May 18, 1970.
S. C. Downs; W. F. Rapp Collection

Warwick, North Dakota **Great Northern**

On the same line as Tolna depot is Warwick station, 79 miles north of Nolan and just south of Devils Lake. Tolna and Warwick depots share a basic design. Warwick was originally a combination depot, but in later years has been used only for freight. This view was made on May 19, 1970.
S. C. Downs; W. F. Rapp Collection

Whitefish, Montana **Great Northern**

A busy sunny day in July, 1969 when this view was taken looking down the wood planked platforms fronting this larger three story station at Whitefish, Montana. The second and third floor exterior walls are stucco with bracing boards, the first floor being mainly wood frame walls. The first floor contains the main station facilities of waiting room, freight rooms, offices. The upper two floors are mainly offices. The building has an interesting white and green combination paint scheme.
Stan F. Styles

Delaware, Ohio **Hocking Valley**
The Delaware station was originally built by the Columbus, Hocking Valley and Toledo Railway in 1890, and is located 24 miles north of Columbus on the C&O's Columbus-Toledo main line. The station still exists however passenger service was discontinued with the coming of Amtrak, but freight services continued until the early 1970's when the station was closed. This older view taken in 1910 shows the station as it originally existed. Today a portion of one end where open covered platforms existed had been removed but the other end remains intact.
Chesapeake & Ohio Historical Society Collection, Courtesy of Mark Camp.

Auburn, Illinois **Illinois Central Gulf**
On the old Gulf, Mobile & Ohio's Chicago-Springfield-St. Louis main line, the Auburn station is 15.5 miles south of Springfield, Illinois. The station joins a large block for passengers to a smaller freight section. The wood-frame station was still in active service when this view was taken on July 1, 1966.
Max Miller

Batesville, Mississippi **Illinois Central Gulf**
The design of this wood-frame depot was a standard one on the former Illinois Central, which is now, due to a 1972 merger, the Illinois Central Gulf. Set in northern Mississippi, the depot is on the north-south main line, 151.6 miles north of Jackson. The depot was still in use, but only for freight, when this view looking down the line was taken on August 19, 1973.
Frank Ardrey

Blue Mountain, Mississippi **Illinois Central Gulf**
Built by the Gulf, Mobile & Ohio as a combination depot, this building now serves freight alone. The depot has been allowed to deteriorate, as seen in this view taken August 19, 1973.
Frank Ardrey

Citronelle, Alabama **Illinois Central Gulf**

On the former Gulf, Mobile & Ohio's Jackson-Mobile main line, Citronelle depot is 32 miles north of Mobile. The viewer is given the impression that the single story and its features such as a corner door and a wide and fancy arched window are about to be crushed by the gigantic roof. The combination depot was still in service when photographed in 1971.
Wilbur C. Thurman Collection

Clinton, Illinois **Illinois Central Gulf**

The station was an important division point, with several of the old Illinois Central's lines crossing through the area. Several operating offices above, as well as station facilities below, were lodged in the big, three-story, brick building. The whole building is divided into three parts, allowing a greater number of windows. Still in active service when this July 6, 1966 view was taken, Clinton station is on the Chicago-Springfield-St. Louis line, 147.3 miles south of Chicago.
Van W. Best

Flora, Mississippi **Illinois Central Gulf**
Built by the Illinois Central Railroad, the small, board-and-batten station is now the home of the Depot Craft Shop. On the Greenwood-Jackson line, 78.6 miles from Greenwood, the building is seen in a view of April 24, 1973.
Frank Ardrey

Newton, Mississippi **Illinois Central Gulf**
This long, brick depot with shallow arches over the doors and windows served both passengers and freight on the Illinois Central. By April 25, 1973, when this photograph was made, only freight was handled here.
Frank Ardrey

Shubuta, Mississippi **Illinois Central Gulf**
Located 96 miles north of Mobile, Alabama, Shubuta station is on the old Gulf, Mobile & Ohio Railroad. A board-and-batten base raises the building considerably off the ground. Originally a combination station, Shubuta now handles freight alone. The depot is still in good condition, as seen in this view of April 23, 1973.
Frank Ardrey

Siloam Springs, Arkansas **Kansas City Southern**

In northwestern Arkansas, close to the Oklahoma state line, Siloam Springs station is on the railroad's Kansas City-New Orleans main line. The single-story building is mostly of brick, but uses stucco on parts of the freight section walls and the trackside gable. The building, still in service, is seen in a view of March 1972.
William F. Rapp, Jr.

Deerfield, Michigan **Lake Shore & Michigan Southern**
Although the structure itself changed little in the years between the two pictures shown here, the rain gutter and leader, the rustic roof covering, the signs, and the feeling of activity that appear in this photograph from around 1910 have all vanished in the later one. In 1881 the LS&MS established a telegraph office at Deerfield station; the first station agent was Cornelius Ward.
Collection of the Roberts-Ingold Memorial Library, Deerfield, Michigan

Deerfield, Michigan **Lake Shore & Michigan Southern**
The wood-frame station at Deerfield still stands on the line between Monroe and Adrian, but is no longer in use. This photograph was taken on March 15, 1975, in the 100th year of the station's existence.
John Uckley

Rhyolite, Nevada **Las Vegas & Tonopah**
The LV&T was completed in Rhyolite in 1906, and the construction of the station began on September 30, 1907 and was completed in June 1908. After little more than ten years, the railroad was abandoned. When this view of May 2, 1969 was taken, this grandiose station of concrete blocks was one of only two buildings occupied in the ghost town of Rhyolite.
Henry E. Bender, Jr.

Beatrice, Alabama **Louisville & Nashville**
Beatrice depot sits at the junction of the L&N line between Corduroy and Selma. Of a standard design, the wood-frame depot is seen in a view made April 20, 1971.
Norman C. Miller, Jr.; Collection of Howard W. Ameling

Beatrice, Alabama **Louisville & Nashville**
This photograph was taken from the freight end of Beatrice depot. It is very similar to the L&N depot in Camden, Alabama. Both were in service when this view was taken in 1971.
Norman C. Miller, Jr.; Collection of Howard W. Ameling

Berea, Kentucky **Louisville & Nashville**
On the Louisville & Nashville's Cincinnati-Knoxville main line, Berea station is 133 miles south of Cincinnati. The brick station is an attractive collection of such features as arched windows, a wide platform shelter, and tile roof. This station was in active service when this view was made on October 11, 1965.
Max Miller

Brewton, Alabama **Louisville & Nashville**
Although still operating as a combination depot, Brewton nowadays handles mostly freight. The dormers on the hip roof act as attic vents. The wood-frame depot is on the Louisville & Nashville's Montgomery-New Orleans main line, 105 miles south of Montgomery.
Wilbur C. Thurman Collection

Camden, Alabama **Louisville & Nashville**
The weathered effects are evident in the Camden depot although many of the colors are also evident such as the dark green roof shingles, lighter green siding with grey on the vertical siding and trim in white. The depot is typical of many that still survive. The Beatrice, Alabama depot has the same color scheme as this depot seen in this view of April 19, 1971.
Norman C. Miller, Jr.

Dickson, Tennessee **Louisville & Nashville**
Directly west of Nashville, the depot is today used for only freight. The vividness of the roof brackets is played against the solidity of the brick walls. The station is well maintained, as seen in this view taken on April 18, 1974.
Frank Ardrey

Greenville, Alabama **Louisville & Nashville**
A brick base supports the windows and cement upper walls; a tile roof tops the whole. On the Louisville & Nashville's Montgomery-New Orleans main line, the single-story building is 44 miles south of Montgomery.
Wilbur C. Thurman Collection

Mascoutah, Illinois **Louisville & Nashville**
Seen here as it looked on its original site in 1968, Mascoutah depot was moved to a park in 1970. The roof extends far beyond the board-and-batten walls on all sides, but especially at this end, where it is supported by two posts. The bargeboards and the shielding that hangs down deeply from between the gables help make this nineteenth-century station especially fine.
National Museum of Transportation, St. Louis, Mo.

Paris, Kentucky **Louisville & Nashville**

The Paris depot was built in the early 1880's of wood frame construction in the late American Victorian style. It was enlarged and modernized in 1904 and 1911 and for many years served as the passenger station for the L&N and the Frankfort & Cincinnati. Presidential candidates made whistle stop speeches at the Paris depot, including Theodore Roosevelt and Thomas Dewey. The depot is no longer used but as a valuable historic landmark the structure is to be preserved and was nominated for the National Register of Historic Places in 1973.

David P. Oroszi

Springfield, Tennessee **Louisville & Nashville**

This compactly, put-together station stands just northwest of Nashville. Originally a passenger station, the wood-frame building has been allowed to deteriorate, as seen in this view taken May 28, 1973.

Frank Ardrey

Lancaster, New Hampshire **Maine Central**
On the Quebec Junction-Beecher Falls line, Lancaster station is 111.4 miles from Portland, Maine. Used mostly for freight service, the Swiss chalet station has a clock tower much like that on the railroad's station at Skowhegan.
Cavalier Collection

Skowhegan, Maine **Maine Central**

The clock tower is similar to, if taller than, that at Lancaster, but the rest of the two stations are very different. Although also touched by the Swiss chalet, Skowhegan station has more of the stiffness associated with Victorian architecture. At the end of the Waterville-Skowhegan branch, the station is 17.7 miles north of Waterville.

Cavalier Collection

)exter, Michigan **Michigan Central**

'he Dexter depot is located 45.4 miles from Detroit on the ld Michigan Central Railroad line, and was built in 886-1887 for passenger service. This view shows the depot s it looked on July 31, 1919 when it was in active service. It ventually was used for freight service for many years but losed in more recent times and converted for use as the)exter Youth Club as it exists today.

:ourtesy C.R. Davidson

Kalamazoo, Michigan **Michigan Central**

This historic plaque is attached to the Kalamazoo station. In the previous picture, the marker can be seen at the right side of the arched door furthest to the right.

John Uckley

.alamazoo, Michigan **Michigan Central**

he work of the distinguished architect, Cyrus L. W. Eidlitz, ıis Romanesque station of brick and stone was built in 886. The Michigan Central, however, had reached Kalama-ɔo much earlier, the first train arriving on February 1, 1846. .t present, most of the trains through Kalamazoo station are .mtrak passenger trains, three westbound and three east-ound, stopping daily. The open shelter at right is connected ɔ a hip-roofed pavilion like that at far left; the once-open ːction that joins it to the central block was bricked up after ıe station's construction.

ɔhn Uckley

Niles, Michigan **Michigan Central**
Built in 1891 for the Michigan Central, this brown, sandstone building with the trackside oriel and a 68-foot tower, bearing a 5-foot clock dial, is a typical example of large-scale Romanesque architecture in North America. The station's line between Detroit and Chicago later belonged to the New York Central, which became the Penn Central. Six Amtrak trains are served daily by the station. The Michigan landmark is seen here in a view of August 1974.
John Uckley

Wimbledon, North Dakota **Midland Continental**
Except for the wood-plank platform and the trackside bay, Wimbledon station might be mistakenly thought to have been built as a private house. The station is on the Midland Continental Railroad's Northern division, which makes connections at Wimbledon with the Soo Line. This view shows the building as it looked on May 18, 1970.
S. C. Downs; W. F. Rapp Collection

Altus, Oklahoma **Missouri-Kansas-Texas**
Built before World War I, the brick station handles only freight today. The structure, thought to have replaced an earlier station, is seen as it looked on July 15, 1966.
Van W. Best; W. F. Rapp Collection

Tulsa, Oklahoma **Missouri-Kansas-Texas**
This M-K-T depot is at one end of the railroad's Muskogee-Tulsa branch, which contains 51.4 miles of track. Believed to be a second-generation depot, it is still active, but now used only for freight. The building, seen in an April 25, 1971 view, is of brick with a tile roof. Lighting fixtures are suspended from the eaves.
Paul H. Dailey; W. F. Rapp Collection

Carthage, Missouri **Missouri Pacific**
Long, stable-like Carthage depot has walls of stone, which also makes up the tower and chimneys. The depot still contains freight facilities and some operating offices, but no longer handles passengers. It is seen in a photograph from October 29, 1970.
William F. Rapp, Jr.

Crete, Nebraska **Missouri Pacific**
This individualistic, stick style station was demolished in 1949. The wide roof was supported by ornate poles. The station stood at the west end of the Auburn-Crete branch and was photographed in October 1947.
William F. Rapp, Jr.

Pleasant Hill, Missouri **Missouri Pacific**
On the St. Louis-Kansas City main line, the station is 33 miles southeast of Kansas City. The brick building with stone trim was erected prior to 1914 and contains a waiting room, large office, and baggage and freight rooms. A brief, full second story at left soon turns into a half-story. The first-floor windows and doors are set beneath shallow arches.
William F. Rapp, Jr.

Gosport, Indiana **Monon**
The New Albany & Salem (Monon) Railroad station at Gosport was built in 1855 and still exists today as a rare example of through-type station. Trains could enter the station building at one end and depart from the other. This early 1970s view shows the station in need of some roof repairs. Although not in use by the railroad today it is a historically significant structure that had become obsolete before the 20th century began. Its future is uncertain.
Mark J. Camp

Monticello, Indiana **Monon Railroad**
On the railroad's Chicago-Indianapolis main line, the depot is 88.4 miles southeast of Chicago. Today the board-and-batten structure is used only for freight and has a Railway Express Agency. This photograph was made in July 1964.
William F. Rapp, Jr.

Chesterton, Indiana **New York Central**
The roof brackets and bay crest are ornate on this Italian station of brick with tile roof. A porte-cochere is attached to one end of the building. Still in use, the station has been well maintained, as seen in this view from April 15, 1968.
William F. Rapp, Jr.

Muncie, Indiana **New York Central**
An early view of the Muncie, Indiana Union Depot which still exists today in good condition. Originally a New York Central depot, later the Penn Central's Cleveland-Indianapolis main line, and used also by Norfolk & Western. The station was closed to passenger service with the coming of Amtrak, although Penn Central still uses it at the time of this writing. Some alterations had been made to the building but the basic structure is still intact and well maintained.
David Oroszi Collection

Canton Junction, Massachusetts
New York, New Haven & Hartford
The stone construction of this station tends to be obstructed by the long canopy in this August 1971 view. Built in 1873, the station is 15 miles south of Boston. The main section has recently been reroofed.
Herbert H. Harwood, Jr.

Franklin, Massachusetts New York, New Haven & Hartford
On the Boston-Franklin line, the depot is 28 miles southwest of Boston. Brick rises to the window sills, above which the walls are stucco. The roof brackets are very elegant. This view was made in August 1973.
Herbert H. Harwood, Jr.

Guilford, Connecticut New York, New Haven & Hartford
Occupied today by a marine supplier, the board-and-batten structure when photographed in July 1974 looked much the same as when built in the 1870s. On the New Haven-New London line, it is 17 miles east of New Haven.
Herbert H. Harwood, Jr.

Kingston, Rhode Island New York, New Haven & Hartford
A spacious, wooden station, Kingston has many attractive details such as the tapered support beneath the side window, the gently arched windows, and the dentils and modillions on the top half-story. On the Westerly-Providence main line, the station is 27 miles south of Providence. The station is still in use, including by long-distance trains between Boston and Washington, when this view was taken in August 1971.
Herbert H. Harwood, Jr.

illis, Massachusetts New York, New Haven & Hartford
his gracious station was built in 1885-1886. Fieldstone akes up the first floor and tower, while the second story is f wood. Originally the first floor held passenger facilities nd the second a library and the town offices. Today town ffices fill the entire building, for the last train from the ation ran in 1965. This trackside view of Millis station was ken in August 1973.
erbert H. Harwood, Jr.

New London, Connecticut

New York, New Haven & Hartford

The large two and a half story New London station is said to be the last station designed by architect Henry H. Richardson, well known for his railroad architectural designs. New London station was built between 1886-1887 in the Romanesque Revival style and was used previously by Central Vermont, New Haven, and Penn Central followed by Amtrak. It is under renovations as seen in this view of August, 1973 and has since been restored and now reused for commercial purposes.
Herbert H. Harwood, Jr.

North Easton, Massachusetts

New York, New Haven & Hartfor

Built in 1881, North Easton station had been designed by th renowned American architect Henry Hobson Richardson, o commission from Frederick L. Ames, whose family owne much of the Old Colony Railroad. In 1882, Ames turned th station over to the Old Colony, that was later to become pa of the New York, New Haven & Hartford. The statio continued in operation until June 1959, after which it wa closed and abandoned for almost ten years. Members of th Ames family in 1969 gave the station to the Easton Historic Society, for use as headquarters and a museum building. Th structure is typical of "Richardsonian Romanesque" with it massive stonework and broad arches. At the rear is porte-cochere. This view shows the building in July 197 when considerable restoration work had been completed.
Herbert H. Harwood, Jr.

Noroton Heights, Connecticut
New York, New Haven & Hartford
A small block sheltered by a large roof, Noroton Heights is being preserved as the last remaining original New Haven suburban station. The platform fence seen in this August 1974 photograph is recent.
Herbert H. Harwood, Jr.

Putnam, Connecticut New York, New Haven & Hartford
Now housing the Grube Camera Shop, the building is in excellent condition and well maintained, as seen in this August 1973 view. Located 26 miles north of Worcester, Massachusetts, the brick station with tile roof is on the Worcester-New London-New York line.
Herbert H. Harwood, Jr.

Thomaston, Connecticut New York, New Haven & Hartford
Ornate panels fill in the roof brackets and the top of the gables on this one-and-a-half-story, brick building. Thomaston is the home of the Seth Thomas clock factory, its main industry. The station is still sound, although, as seen in this July 1974 view, repairs are in order.
Herbert H. Harwood, Jr.

Wallingford, Connecticut New York, New Haven & Hartford
This sturdy, Victorian structure is of brick with a mansard roof. The brackets supporting the cornice are echoed beneath the dormer arches. Located 43 miles from New Haven, the station is seen in a view of July 1974.
Herbert H. Harwood, Jr.

Walpole, Massachusetts New York, New Haven & Hartford
A two-story bay sends out two wings at a right angle to fit alongside a track intersection. The tower is unusual in design, and the ornament along the peak of the roof should not be overlooked. On the Boston-Franklin line, the station is 19.5 miles southwest of Boston. The photograph was made in July 1970.
Herbert H. Harwood, Jr.

Westerly, Rhode Island **New York, New Haven & Hartford**
A very handsome, Italian station of brick and stucco with tile roof has attached to each end a pavilion, one open and one enclosed. None of the materials used in the station are especially expensive, yet it looked grand indeed. Good design can create much beauty at not much cost. The station, near the southern boundary of Rhode Island, is on the New Haven-Boston line. Still active, it is in excellent condition in this view of August 1971.
Herbert H. Harwood, Jr.

Windsor, Connecticut **New York, New Haven & Hartford**
On the Hartford-Springfield line, the station is 19 miles south of Springfield. Built in the 1870s, the building is a simplified version of the station at Wallingford. The station is closed and in a state of disrepair as seen in this view of July 1974.
Herbert H. Harwood, Jr.

Windsor Locks, Connecticut

New York, New Haven & Hartford

The plain, brick building has a fancy roof with scalloped shingles and highly decorated gable eaves. The peak of the dormers meet that of the roof; the dormers seem all the further up the roof for its trackside overhang. The station has been placed on the State Inventory, State Historical Preservation Plan. Rail-auto service was once available from this point. The station is shown in a view from July 1974.
Herbert H. Harwood, Jr.

New Bern, North Carolina **Norfolk Southern**

Located in the southwest part of the state, New Bern depot serves the Norfolk Southern, the Seaboard Coast Line, and the Southern (A&EC) railroads. However, only freight is handled at the depot. This view of July 12, 1974 was taken from the Norfolk Southern side.
Howard W. Ameling

New Bern, North Carolina **Norfolk Southern**
Austerity is an asset to the design of this symmetrical building. Taken on the same day as the preceding picture, this photograph shows the same facade, but from the other end.
Howard W. Ameling

Berryville, Virginia **Norfolk & Western**
Although nearly identical to the railroad's station at Shepherdstown, West Virginia, the station is of a lighter brick. At trackside, the semi-octagonal bay of the brick first floor supports a half-timbered second story that emerges from the hip roof. The high freight platforms have disappeared, but not without a trace, for their outlines, particularly those of the steps on the side wall, remain. The station was used for freight-only service when this view was taken in May 1971.
Herbert H. Harwood, Jr.

Saint Marys, Ohio **Norfolk & Western**
Saint Marys depot is on the Sandusky-Frankfort line, 110 miles from Sandusky, and on the Saint Marys-Minster branch line, 9.9 miles north of Minster. The depot has ornate window and door pediments and roof brackets. It handled only freight when this July 21, 1971 photograph was made.
Howard W. Ameling

Shepherdstown, West Virginia **Norfolk & Western**
Very similar to the Berryville, Virginia station of the railroad, this station is of darker brick and still has the high freight platform in place at the end. Originally used for passenger service, the station was used for freight exclusively when this April 1975 view was made.
Wilbur C. Thurman Collection

Waverly, Virginia **Norfolk & Western**
Located 61.7 miles from Norfolk, this brick station is in excellent condition as seen in this view from July 11, 1974. Note that the fire wall, dividing the passenger and freight portions of the building, extends above the roof.
Howard W. Ameling

Auburn, Washington **Northern Pacific**
The Northern Pacific shares this still-active station, 19 miles from Tacoma, with the Great Northern. A row of zigzag shingles separate the square shingled upper walls from the siding of the lower. The symmetry of the roof, which descends in three sections, and the looser symmetry of the first floor, have been disturbed by the addition to the freight facilities seen in the foreground of this September 10, 1968 photograph.
Henry E. Bender, Jr.

Chehalis, Washington **Northern Pacific**
Chehalis depot, which has been used jointly with the Great Northern, was built about 1912-1913 by the Northern Pacific, now the Burlington Northern. The station's past is linked to Presidents William McKinley, Theodore Roosevelt, and William Howard Taft, and the building, unoccupied but in good condition when this view was made in October 1968, is listed on the National Register of Historic Places.
Stan F. Styles

Lewiston, Idaho **Northern Pacific**
This large, but rather strict, station of brick was built in 1909 by the Camas Prairie Railroad and was still in service when this photograph was taken in August 1973. Owned and used jointly by the Northern Pacific and Union Pacific railroads, the station is in the northwest part of the state, close to the Washington state line.
Stan F. Styles

Sandpoint, Idaho **Northern Pacific**

The Sandpoint station was intended to have a character distinctive from that of the stations in other Idaho small towns. It is believed the state's only Gothic station, but is just barely Gothic, possessed of such features as the arches of the first floor windows, that give only the faintest suggestion of the style. Two other, perhaps more significant facts pertain to the station: its name and that of its location were given by the town's principal developer, the Northern Pacific Railroad, and it is the only station in Idaho still used for passenger service after the advent of Amtrak. The photograph shows the station as it appeared in July 1968.

Stan F. Styles

Fort Seward, California **Northwestern Pacific**
A Budd car with the markings of the Northwestern Pacific's parent, Southern Pacific, stops to discharge mail and passengers at Fort Seward, once an army outpost and now an infrequently active lumber depot. The two-story station is of brick on the first floor and half-timbering on the second, that extends forward at trackside to rest on brick columns and form a colonnade.
Redwood Empire Association

'ort Seward, California **Northwestern Pacific**
he station is seen.as it was on March 20, 1971, just twelve ays before its conversion from passenger-and-freight service) freight-only, with the coming of Amtrak. Passenger service 'as brisk at this time, as evidenced by the fully loaded RDC 'ain, the "Redwood," as it paused in this scenic location.
'enry E. Bender, Jr.

Petaluma, California **Northwestern Pacific**

The Northwestern Pacific's depot at Petaluma is on the Southern Pacific's coast line, north of San Francisco. The neat and trim, little depot is in the Spanish Mission style and has a red tile roof and stucco walls. This streetside view was taken on March 20, 1971.
Henry E. Bender, Jr.

Willits, California **Northwestern Pacific**

This fascinating all-redwood station is on the Northwestern Pacific's Eureka-Willits-San Francisco line, 139.5 miles north of San Francisco. The station opened in 1916 and served both the Northwestern Pacific and California Western railroads. The station is today used by freight crews of both railroads and is still a stop for the CWR "Skunk" trains. This view shows the station as it looked on October 13, 1968.
Henry E. Bender, Jr.

Middle River, Maryland **Penn Central**

Located on the Maryland Division this former Pennsylvania Railroad structure is now the Penn Central Company Middle River Freight Station as seen in this view of August, 1970. The station has interesting framing style on the second floor with dormers breaking the roof pattern at various points. In contrast to the lower first floor brick styling, the second floor looks much like it was placed atop the first floor to combine two separate buildings into one.
Herbert H. Harwood, Jr.

Berwyn, Pennsylvania **Pennsylvania Railroad**

This slightly Gothicized, brick building stands on lowered ground, while the roadway to the right has been raised and the two heights are connected by stairs, a sample of the maneuvering necessary to separate track and street levels in suburban settings, such as this one, outside Philadelphia. The silverliner easing into the well maintained station on a cloudy day in 1974 belongs to the Southeastern Pennsylvania Transportation Authority, which took over most of the area's public transit from a number of troubled companies.
Bob Storks

Bridgeville, Delaware **Pennsylvania Railroad**
On the Wilmington-Cape Charles-Portsmouth line, the depot is 77 miles from Wilmington. This plain, brick building is used only for freight and, in this December 1970 photograph, appears well maintained.
Herbert H. Harwood, Jr.

Clayton, Delaware **Pennsylvania Railroad**
This nineteenth-century, brick station features semicircular arches over windows and doors. The customary trackside bay is absent. Seen in a recent view, the station is on the Delaware division.
Photo by Herbert H. Harwood, Jr.; W. F. Rapp Collection

Decatur, Indiana **Pennsylvania Railroad**
Built of brick with limestone trim, this decorative, little station is in the Tudor style. Special features are the gable's air vents and the small projecting roofs supported by curving brackets above the doors at either side of the trackside bay. This view shows the depot as it looked on February 20, 1972.
David P. Oroszi

Federalsburg, Maryland **Pennsylvania Railroad**
On the Seaford-Cambridge line, this depot is 9.8 miles from Seaford. This small, wood-frame depot was intended for freight-only service and built with a freight room and offices. The photograph shows the depot in December 1970.
Herbert H. Harwood, Jr.

Greensburg, Pennsylvania **Pennsylvania Railroad**
This beautiful station stands 30.8 miles from Pittsburgh. The Tudor-Jacobean structure includes porte-cochere and tower, and is rich in architectural embellishments. The station had passenger and freight facilities and is still structurally sound. All the same, it was bypassed by Amtrak and left vacant. The view shows the station as it looked in June 1971.
Herbert H. Harwood, Jr.

Greenwood, Delaware **Pennsylvania Railroad**
Used only for freight, this wood-frame structure of standard design is seen in a December 1970 view. The depot is on the Wilmington-Cape Charles-Portsmouth line, 72.5 miles from Wilmington.
Herbert H. Harwood, Jr.

Martinsburg, West Virginia **Pennsylvania Railroad**

Although the one-story structure at right is still maintained as a freight depot, the two-floor passenger station is now the Perfection Outlet Store. The bricks that make up the first story form a lively pattern between the window sills. At the same level on the second story, the shingles change from square to scalloped. On either side of the bay, small roofs between floors make trackside shelters. The buildings are seen in a June 1974 photograph.

David P. Oroszi

Milton, Pennsylvania **Pennsylvania Railroad**

Ornate brickwork distinguishes this station, 66 miles from Hamburg, as does the fancy ironwork on the platform canopy. While still sound, the station is closed, as seen in this March 1975 view.

Wilbur C. Thurman Collection

Moorestown, New Jersey **Pennsylvania Railroad**
The Moorestown station as it looked September 2, 1952. Looking west, this view gives an indication of the large windows in comparison to the doors. The series of side roof braces dominate the structure but at the same time adds to its interesting overall style and design.
Edward W. Weber

Moorestown, New Jersey **Pennsylvania Railroad**
Moorestown passenger station is 10.4 miles from Camden in New Jersey on the Pennsylvania Railroad's Camden-Pemberton line. The wood frame board and batton constructed station features arched windows and doors, ornate carved brackets and roof overhang braces blended in a harmoneous design. This most interesting station is no longer used as a station facility but is used now as the Depot Antiques Shop as pictured in this view of February, 1973.
Herbert H. Harwood, Jr.

Moylan-Rose Valley, Pennsylvania **Pennsylvania Railroad**
Still serving as a commuter station on the Media line, 13 miles from Philadelphia, the building is pictured in June 1974. From trackside can be seen the small baggage room at platform level, the porch at the higher level of the waiting room, and the enlivening roof brackets.
Bob Storks

Newark, Delaware **Pennsylvania Railroad**
This successfully Gothic, brick building features bands of darker color brick and of brick set in patterns, mildly fancy bargeboards, and suitably shaped windows. A bridge lifts the street above the track grade, as seen in this view from June 1972.
Herbert H. Harwood, Jr.

New Hope, Pennsylvania **Philadelphia & Reading**
The New Hope depot in its original location as seen in this view of August 1, 1952 looking westbound. A notice of abandon passenger service is on the end door. Some time after this view was taken, the depot was moved back closer to the freight station as seen in the accompanying view of October, 1970. Note there are no steps to the doors, the depot being on grade level. The roof trim and rainwater leaderboard have since been removed.
Edward W. Weber

New Hope, Pennsylvania **Philadelphia & Reading**
The New Hope depot built in 1891, now the property of the New Hope & Iveland Railroad, has been restored and serves as a terminal for short steam excursions. The view shows how it appeared in October 1970.
Herbert H. Harwood, Jr.

Strafford, Pennsylvania **Pennsylvania Railroad**
The upper story of this station consists of a building erected for the Centennial Exposition, held in 1876 in Philadelphia's Fairmount Park. After the fair, the structure served from 1885 to 1886 as the station at Wayne, one of the communities along the chain of Philadelphia suburbs known as the Main Line, through which runs the old Pennsylvania Railroad's tracks to Paoli, then west to Pittsburgh. From Wayne, the station moved briefly to Eagle, now Devon, nearby on the same line, then finally in 1887 came to rest at Strafford, the next stop closer to Philadelphia. The board-and-batten first story was added to raise the station to the height of the tracks, which are above street level. Seen in this view of September 1971, the station operates still as a commuter stop and is in fine condition.
Herbert H. Harwood, Jr.

Wycombe, Pennsylvania **Philadelphia & Reading**
The Wycombe depot was built in 1891 and was originally called Walton. This earlier view shows the depot when still in active service on the Reading Railroad and located on a branch line in east Pennsylvania. The depot originally contained a general waiting room, an office in the circular tower portion, and a freight room that occupied about half the overall floor space of the building.
Edwin P. Alexander

Wycombe, Pennsylvania **Philadelphia & Reading**
The Wycombe depot as it looked in September, 1971 is now owned by the New Hope & Ivyland Railroad who purchased this segment of the line. The depot still exists and is used mainly for storage and as an office by the current owners who also provide summer tourist live steam operations.
Herbert H. Harwood, Jr.

Wynnewood, Pennsylvania **Pennsylvania Railroad**

During the years following the Civil War, a number of such stations of rough stone were erected in Philadelphia neighborhoods and suburbs. Some of this hardy breed of stations, like the Wynnewood station built in the 1870s on the Philadelphia-Paoli line, still serve commuters. Note that the platform shelter is supported by poles in pairs, unlike later, more sophisticated umbrella canopies upheld by single columns.
Bob Storks

Gettysburg, Pennsylvania **Reading Railroad**

The stick style station, with a tower that forms a tiny second story, is seen newly painted and being readied for reuse by a commercial business, other than a railroad. Originally a combination station but later handling solely freight, the building is pictured in March 1973.
Herbert H. Harwood, Jr.

Leesport, Pennsylvania **Reading Railroad**
On the Philadelphia-Reading-Pottsville line, this structure is 7.5 miles west of Reading. The small, brick depot once served both passengers and freight, but is no longer used by the railroad; a scrap dealer's office now occupies the building. Except for the bay and freight platform, it does not look all that much like a station, and it should be noted that the exterior passenger shelter is at one end rather than trackside.
Herbert H. Harwood, Jr.

Sinking Springs, Pennsylvania **Reading Railroad**
Here is an unusual, little station, as seen from roof, bay, and overall shape. What a shame that it is boarded up and in disrepair, as this August 1972 photograph shows.
Herbert H. Harwood, Jr.

Winterthur, Delaware **Reading Railroad**
North of Wilmington and near the Pennsylvania state line is this 1908 building, erected as a privately owned station for a duPont estate at Winterthur. Covered by scalloped shingles but with shingle variations over windows, the structure is now occupied as a residence as seen in this view of August 1972.
Herbert H. Harwood, Jr.

Wissahickon, Philadelphia, Pennsylvania **Reading Railroad**
The Reading erected numerous versions of a brick and half-timbered station, of which this 1892 building is one. Of special interest here are the trackside overhang supports, in which large, curved brackets spring from low on notched piers. Some of the windows of this still-active commuter station have been closed up, as seen in this February 1974 view.
Herbert H. Harwood, Jr.

Mancos, Colorado **Rio Grande Southern**
Clapboards cover the first floor, shingles the second, at Mancos, a two-story station. Such stations often have living quarters above. The Galloping Goose rail car No. 5 has drawn up alongside the station in August 1951.
Ernest S. Peyton

Ridgway, Colorado **Rio Grande Southern**
Similar to Mancos station, Ridgway is a combination station with a two-story bay at trackside. It is still August 1951, and this time Galloping Goose rail car No. 4 has arrived.
Ernest S. Peyton

North Bennington, Vermont **Rutland**
The North Bennington station was built in 1880 and replaced an older wood frame building on the same site. The station served as the gateway to the village for over half a century until the mid 1930's when Rutland's passenger service declined. The station fell into disrepair and by mid century it was a derelict but still a reminder of the areas important railroad past. In 1971 the station was given new life in being converted to municipal offices resulting in the preservation of this fine structure which was nominated for the National Register of Historic Places in 1973.
Collection of C.L. Andrews

Clinton, South Carolina **Seaboard Coast Line**
The wood-frame, combination depot has a long freight section. On the New York-Birmingham line, 402 miles south of New York City, the depot was built by the Seaboard Air Line, now the Seaboard Coast Line. The structure appears to have been extensively remodeled. It is still in service in this June 27, 1966 photograph.
Max Miller

Delray Beach, Florida **Seaboard Coast Line**
The Mediterranean station at Delray Beach still has an agent-operator, but was totally assigned to Amtrak duties when this photograph was taken on June 16, 1973. On the West Palm Beach-Miami line, the station is 52.5 miles north of Miami.
Henry E. Bender, Jr.

Hoffman, North Carolina **Seaboard Coast Line**
Seen on its original site in a January 18, 1975 view, this station was sold, and in May 1975 was moved from the property. The wood-frame structure, with metal shingle roof, was built in 1908 as a combination (passenger-and-freight) station. The identifying sign is one of the few that tell not only the community but also the state.
R. D. Patton

Naples, Florida **Seaboard Coast Line**

L. Philips Clarke was the architect of this Mediterranean station, built in 1926-1927 by the Seaboard Air Line, now the Seaboard Coast Line. The town of Naples was incorporated in 1925, so the station is one of the oldest structures in the community and also one of the most important to its earliest history. The first train arrived at the station on January 27, 1927, even though the building was still under construction. This trackside view has the passenger section in the foreground; the freight station is behind the raised platform, beside which terminates a short freight spur.

Seaboard Coast Line

Naples, Florida **Seaboard Coast Line**

In this photograph, the other side of the station is seen from the passenger end. The Seaboard Air Line, in 1942, discontinued regular service to Naples and sold the property to the Atlantic Coast Line and other railroad companies. The ACL merged with the SAL in 1967 and now operates as the Seaboard Coast Line. The station is now used as a freight warehouse and quarters for an historical association. The building was placed on the National Register of Historic Places in 1974.

Seaboard Coast Line

Thomasville, Georgia **Seaboard Coast Line**
Built by the Atlantic Coast Line, now the Seaboard Coast Line, Thomasville station is seen in a view from the 1920s. The symmetrical building is in bricks of alternating shade with light brick above windows and doors and in the stringcourse at the top of the first story. The picture identifies the section in the foreground as a restaurant. Station facilities were on the first floor of the station, and offices on the second.
Seaboard Coast Line

Woodland, North Carolina **Seaboard Coast Line**
The small, wood-frame depot, still in active service when this June 1964 view was taken, is used for freight-only service now. Built by the Atlantic Coast Line, now the Seaboard Coast Line, Woodland depot is on the Boykins-Lewiston line, 19 miles from Boykins. The state, as well as the town, is given on the sign.
Max Miller

Minot, North Dakota **Soo Line**

Designed by William Kenyon, this one-and-a-half story, brick passenger station was built in 1913-1914. Still in active service when this view was taken in August 1968, the station also houses a post office and Western Union office. The high trackside bay is of unusual form, as are the dormer windows. The brickwork is very decorative, particularly on the end gables and porte-cochere.
Stan F. Styles

Austell, Georgia **Southern Railway**

Separating the office and freight sections is a fire wall that projects above the roof and follows its contour. A narrow, concrete-block platform extends from the trackside freight door to a siding beside the through track. At the junction of the Birmingham and Cariotte divisions, the station is seen in a view of March 19, 1970.
William F. Rapp, Jr.

Camden, South Carolina **Southern Railway**
On the Kingsville, Rock Hill, and Marion line, the depot is 37 miles north of Kingsville. The wood-frame building looks almost residential, especially with its classical doorway (probably an addition). A concrete-block base supports the structure. Used for freight-only service, the depot was still in use when this October 1973 view was taken.
Frank Ardrey

Lexington, Kentucky **Southern Railway**
Suitable to an area that brings thoroughbred horses to mind, this elegant, two-story station with curved bays and a portico, supporting a balcony behind colossal columns, speaks of Southern graciousness. On the railroad's main line, 79.4 miles south of Cincinnati, the station is on the north-south Cincinnati-Chattanooga line and the east-west Lexington-St. Louis line. The station was still in active service on July 4, 1962 when this view was taken.
William F. Rapp, Jr.

Linden, Virginia **Southern Railway**
Linden depot is at the summit of Manassas Gap, on the Manassas-Harrisonburg branch line. The rustic building has its doors and windows set deep in its thick walls, and a porch not, just at trackside but on either side. This view is from May 1971.
Herbert H. Harwood, Jr.

Mableton, Georgia **Southern Railway**
Now handling only freight, the depot is in good condition, as seen from this view of March 19, 1970. The clapboard structure, of familiar design, has at the freight end a high platform to facilitate loading.
William F. Rapp, Jr.

The Plains, Virginia **Southern Railway**

This brick depot adheres to a customary plan, but derives special interest from its tile roof, semi-octagonal bay, and latticed windows. On the Mediterranean-Harrisonburg branch line, The Plains depot is 19.8 miles west of Manassas. The building was still in use when this view was taken in May 1971.

Herbert H. Harwood, Jr.

Goodyear, California **Southern Pacific**
Built in 1891, the old Goodyear depot is seen in a 1913 view. Once the wood-frame building contained passenger facilities and a Wells Fargo express agent. In 1930, a new, slightly larger station in the Spanish Mission style was built near this site to handle all the passenger traffic. The depot, pictured, continued to serve as a freight station and express office until it was abandoned in 1939.
Southern Pacific

Alturas, California **Southern Pacific**
Of stone, said to be pink-tinged, and with wood-shingle roof, he station was built by the Nevada-California-Oregon, a narrow-gauge railroad. About 1910, the California Railroad Commission decided that the station was inconveniently located and told the N-C-O to move it closer to town. The ailroad numbered all the stones, sawed the roof in half, and ransplanted the station a mile or so southward, to the present site. Later the station was acquired by the Southern Pacific, which changed the gauge to standard. No longer in ailroad service, the building is now used by the Alturas garden club.
Thomas Armstrong

Fernley, Nevada **Southern Pacific**

A standard, Southern Pacific combination (passenger-and-freight) depot, the wood-frame building contained living quarters above for the agent, as did many such stations. The depot was still in use when this December 24, 1968 view was taken.

Van W. Best; W. F. Rapp Collection

Loomis, California **Southern Pacific**
Seen in 1913, this station is a variation on Southern Pacific Combination Station Design No. 23. This design was for a wood-frame building with station agent's living quarters on the second story. At this station, a platform at the level of both freight and passenger cars runs the length of the station and more.
Southern Pacific

.akeview, Oregon **Southern Pacific**
)riginally erected by the Nevada-California-Oregon Railroad etween 1912 and 1915, the station was acquired and is still sed by the Southern Pacific. Classical details are done oldly, and rough stone is used for springers, keystones, and second-story stringcourse. The building is still sound, as een in this 1973 photograph.
homas Armstrong

Luning, Nevada **Southern Pacific**

The depot was used for freight-only service when this view was taken on December 24, 1968. An air conditioner has lately been mounted on the roof for the comfort of the second-floor occupants.

Van W. Best; W. F. Rapp Collection

Monmouth, Oregon **Southern Pacific**

The station was on the Southern Pacific's Airlee branch between Dallas and Airlee, both in Oregon. Construction of the line was first undertaken by the Oregonian Railway Company, on September 17, 1881. The line did not reach Monmouth until 1905, apparently the approximate time this depot was built.

Southern Pacific

Monmouth, Oregon **Southern Pacific**
The original depot was superseded in 1914 when this new depot was built. The replacement building, however, was abandoned when the line itself was abandoned in 1934.
Southern Pacific

San Carlos, California **Southern Pacific**
The San Carlos station was built in 1906 of sandstone construction giving it a sturdy but attractive design. The station still exists though leased as a real estate office. San Carlos remains as a stop on the Southern Pacific Peninsula commuter line, however, the agency has been discontinued and commuters now use an adjacent shelter. This view, taken on May 31, 1969 shows San Carlos in its new reused occupancy of The Avenue Realty.
Henry E. Bender, Jr.

Santa Barbara, California **Southern Pacific**

Santa Barbara station, with its two stories and fine decoration, is an example of the Spanish Mission style, grown elaborate during the glory days of railroading. The first floor arcade forms a trackside shelter, showing how willed to railroad architecture was the style.
Southern Pacific

Stockton, California **Southern Pacific**
Here is a pre-Amtrak view of Stockton station taken on June 21, 1969, when the "Sacramento Daylight" still stopped here twice daily, once going each way. The sturdy, brick building with tile roof is in excellent condition in this picture. The antefixes around the trackside marquee, the ornate window frames, and the band of decoration below the main block's cornice are wondrous to behold on a sunny California day.
Henry E. Bender, Jr.

Terra Bella, California **Southern Pacific**
The usual, stucco finish is omitted from the brick, Spanish Mission station with red tile roof. Still in fine condition as seen in this April 30, 1971 view, the building is now an office for a water well drilling company.
Henry E. Bender, Jr.

Toledo, Oregon **Southern Pacific**
Picturesque locomotive No. 7 has stopped at Toledo station, and the train and station crews, as well as some local visitors, have posed for this turn-of-the-century view. The brick, combination depot with wood-plank platform is on a line originally the Carvallis & Eastern, but now the Southern Pacific's West End branch.
Southern Pacific

Toledo, Oregon **Southern Pacific**
Southern Pacific's replacement, combination depot at Toledo is no longer used for passengers. A concrete-block base supports stucco upper walls on the single-story depot. The transverse roofs have gables at one end only and are hip at the other, where they join the pitch of the long, lower roof. This view shows the depot as it looked in June 1960.
Stan F. Styles

Bend, Oregon **Spokane, Portland & Seattle**
Built about 1911 by the Oregon Trunk Railroad, the depot is now owned jointly and well maintained by the Union Pacific and the Spokane, Portland & Seattle railroads. Doors and windows are set deep in the fine, massive, stone walls, on which rests a low hip roof. Passenger and freight facilities were both in use when this view of September 4, 1968 was taken.
Henry E. Bender, Jr.

Madras, Oregon **Spokane, Portland & Seattle**
The lower slope of the gambrel roof above the passenger facilities would not allow for much of an exterior shelter, unless the second story were widened far more than here. The freight section has a gable roof of equal width. The Oregon Trunk, owned by the Spokane, Portland & Seattle and Union Pacific railroads, built the station. Southeast of Portland, it is near Mt. Hood and at an elevation of 2,458 feet.
Henry E. Bender, Jr.

Washtucna, Washington **Spokane, Portland & Seattle**
On the SP&S main line, Washtucna depot is 88.6 miles from Spokane. The single-story, wood-frame building in horizontal siding is a combination station, but was used mainly for freight when this picture was taken in July 1973.
Stan F. Styles

Fayetteville, Arkansas **St. Louis-San Francisco Railway**
This single-story, Mediterranean station has brick exterior wainscoting below stucco upper walls. An open shelter is at one end, and note that there are two trackside bays instead of the usual one. In northwest Arkansas, the building is on the railroad's north-south line, 63 miles north of Fort Smith. The station was still in use when this March 17, 1972 view was made.
William F. Rapp, Jr.

Miami, Oklahoma **St. Louis-San Francisco Railway**
Still sturdy and in active service when this October 30, 1970 view was taken, the building was erected by the Kansas, Oklahoma & Gulf Railway and later became the property of the St. Louis-San Francisco Railway. Freight facilities, as well as division offices, were lodged in the depot.
William F. Rapp, Jr.

Springdale, Arkansas **St. Louis-San Francisco Railway**
On the Frisco's St. Louis-Fort Smith-Paris line, the station remained in use when this view was made on March 17, 1972. The brick is a lighter shade below the level of the window sills than above. The station handled both passengers and freight.
William F. Rapp, Jr.

Wilson, Arkansas **St. Louis-San Francisco Railway**
The long, narrow, wood-frame depot was intended as a combination station, but only accommodates freight today. Near the Arkansas state line, the depot is on the St. Louis-Memphis main line, 40.6 miles north of Memphis. A Western Union Telegraph and Cable Office is also housed here, as seen in this view of May 1966.
Max Miller

Arden, Nevada **Union Pacific**
On the Salt Lake City-Los Angeles main line, little Arden depot is 11 miles southwest of Las Vegas and near the California state line. The wood-frame building also contains—at right in this view—the Arden post office.
William F. Rapp, Jr.

Ashton, Idaho **Union Pacific**
In the last days of Union Pacific passenger service, Ashton was the terminus of the Yellowstone branch. Even in earlier times, the depot was the end of the line when the park was closed. Located 51 miles from Idaho Falls, the depot is pictured in August 1965.
William F. Rapp, Jr.

Butte, Montana **Union Pacific**
At the northern end of the Idaho Falls-Butte branch line of the Union Pacific, Butte station is 212 miles north of Idaho Falls. The big, red, brick building was headquarters of the railroad's Montana operation, and is used jointly by the Northern Pacific Railroad. The station was in active service when this August 1973 photograph was made.
William F. Rapp, Jr.

Columbus, Nebraska **Union Pacific**
The Columbus, Nebraska station was built in 1887 by the Union Pacific Railroad Company and is located 82 miles west of Omaha. The elegant cut stone passenger station was expanded in 1909 by a 30 x 67 foot addition blended skillfully to match the existing structure. The Union Pacific plans to improve the station which is an important link to the development of Columbus in early days. This view shows the station as it looked in June 1962.
William F. Rapp, Jr.

Cozad, Nebraska **Union Pacific**
The Cozad station was erected in 1925-1926 period as a passenger station for Union Pacific and having limited freight handling facilities. The station still exists and is located in Dawson County, south central Nebraska close to the Platts River. Interior rooms include a waiting room, agent's room, several smaller rooms and a baggage room at one end. The building is constructed of brick up to the window sills with stucco above this up to the roof lines. This view from the west shows the station as it looked on August 21, 1974.
William F. Rapp, Jr.

Kimball, Nebraska **Union Pacific**
The semi-octagonal bay of this wood-frame, combination depot is itself against a flat bay that projects slightly from the trackside wall. The depot was still in use when this view was taken on July 18, 1967. On the Union Pacific's main line, the depot is 65 miles from Cheyenne, Wyoming.
Howard W. Ameling

Logan, Utah **Union Pacific**
The Logan station was originally owned by the Oregon Short Line Railroad, now Union Pacific Railroad and was built in 1898 during Oregon Short Line ownership. This more recent photograph shows the station is still well maintained and in use by the Union Pacific Railroad but only for ticket sales and as an express office. The cut grey stone work contrasts with the smoother dark red brick, with the long hip roof sheltering 178 feet of the platform along track side.
Herbert H. Harwood, Jr.

Montpelier, Idaho **Union Pacific**

Reduced to basic shapes, the brick station has the roof brackets eliminated and the window sills and lintels rendered as plain blocks. In southeast Idaho, the building is on the Union Pacific's main line, 99 miles east of Pocatello. The station was still in use when this photograph was made in August 1965.
William F. Rapp, Jr.

Ovid, Colorado **Union Pacific**

A small-town depot in northeast Colorado, Ovid is on the Union Pacific main line. The brick exterior wainscoting that rises to the stucco upper walls drops slightly at the trackside bay. Containing a Railway Express Agency and a Western Union office, the depot, still in active service, is well maintained, as seen in this view of October 22, 1967.
William F. Rapp, Jr.

Park City, Utah **Union Pacific**

Victorian architecture is at its jolliest in the porches of this station, just northeast of Salt Lake City. The first-floor porch, now partially filled in by a pavilion, bearing a Railway Express Agency sign, is rectangular with trelliswork between the top of its piers. Its lines lead up to the second-story loggia, where the design explodes into a scallop-shingle back wall, a semicircular arch surrounded by trelliswork, and a railing filled in by a vertical, wavy pattern at center with a sunburst at each side. Used mainly for freight nowadays, the station is seen in this view of January 2, 1975.
Henry E. Bender, Jr.

Pine Bluffs, Wyoming **Union Pacific**
The raised freight room floor is reached via a sloping platform, which is set upon a grade that tilts in the opposite direction. Still in use when this view of July 18, 1967 was taken, the depot has a Railway Express Agency and a Western Union office. On the Nebraska state boundary, the depot is on the east-west main line of the Union Pacific, 43 miles from Cheyenne.
Howard W. Ameling

Riverside, California **Union Pacific**
Construction of the Riverside station was completed in 1904 by the San Pedro, Los Angeles and Salt Lake Railroad. This view taken August 10, 1967 shows the station when it was still in active passenger service which ceased on May 1, 1971. The station has since been closed but Riverside city officials plan to preserve the building as a cultural heritage landmark.
Max Miller

Solomon, Kansas **Union Pacific**
Of Victorian influenced design, Solomon depot is on the National Register of Historic Places mainly for architectural reasons. It is one of the best remaining examples of railroad Victorian architecture designed in stone. It is also a fine example of the type of permanent structure preferred in Kansas when railroads were an important part of the development of Kansas and the west in the late 19th century.
William F. Rapp, Jr.

South Torrington, Wyoming **Union Pacific**
Built in the mid-1920s, South Torrington station looks almost like a row of village shops. A brick base supports stucco upper walls, and bricks are set at corners in quoin-like arrangements. Still in use, the station is on the Union Pacific's North Platte cutoff. The building is well maintained, as seen in this June 18, 1967 view.
William F. Rapp, Jr.

West Yellowstone, Montana **Union Pacific**
With rough stone upon its walls, piers, and chimneys, this station was suitable for the Union Pacific's entrance to Yellowstone National Park. Thousands of visitors have passed through this station. Though seemingly of sturdy construction, the station was damaged by an earthquake in 1959 and abandoned by the railroad. This view from August 1965 shows it with windows and doors boarded up.
William F. Rapp, Jr.

Moravia, Iowa **Wabash Railroad**
Two patterns of board-and-batten siding, plus vertical, horizontal, and diagonal timbering, plus double rows of scallop shingles alternating with double rows of zigzag shingles, coat this station and its high forehead. At the left are two different kinds of baggage doors; perhaps the one at the far left was an addition. The view is of June 2, 1957.
Howard W. Ameling

Reno, Nevada **Western Pacific**

Later acquired by the Western Pacific, this building was erected between 1912 and 1915 by the Nevada-California-Oregon Railroad, to replace another large building a few blocks away. Originally used as a station with general offices, the two-story, brick structure with stone trim and tile roof is still occupied by the railroad, though part of it is rented out as the main office of the Sierra Wine & Liquor Co. The building is in excellent condition, as seen in this view of 1973.

Thomas Armstrong

Stockton, California **Western Pacific**
Stockton station is seen on June 21, 1969 when it still offered passenger service. The building, in the Spanish Mission style, was boarded up and advertised for sale soon after the last run of the "California Zephyr" on March 22, 1970.
Henry E. Bender, Jr.

Bowerston, Ohio **Wheeling & Lake Erie**
Shown in a rare view, circa 1890, this eminently, Victorian station has a number of unusual elements. The passenger end, at right, has delicate siding that runs in several directions, while the freight end, at left, has plain board-and-batten walls.
Howard W. Ameling Collection

Cleveland, Ohio **Wheeling & Lake Erie**
The station appears to be a two-story, Elizabethan building, as seen in this rare view from streetside. However, there are two brick stories beneath, that rise from track level to support the upper part of the structure.
W&LE, Howard W. Ameling Collection

Dalton, Ohio **Wheeling & Lake Erie**
The old Dalton depot as seen in this July 1970 view was once quite active although long since out of service as evident from the rusty rails and weeds growing along the tracks. A wood platform had once fronted the depot with steps leading down along side the building to the lower rear ground level. The wood frame structure is believed to be no longer existing since this view was taken.
Herbert H. Harwood, Jr.

Dundee, Ohio **Wheeling & Lake Erie**
A row of wooden darts surround the upper walls of this one-story depot. The view is believed to have been made before 1918. The last time a passenger train made a stop here was in 1932, although light freights continued to use the depot. The paint scheme had been altered to brighter colors many years after this view was taken.
Howard W. Ameling Collection

Fremont, Ohio **Wheeling & Lake Erie**
Also known as the "Palmer House," this station was built about 1850, and is seen in a view from circa 1895. The structure was replaced around the turn of the century by a smaller, brick station.
Howard W. Ameling Collection

Fremont, Ohio **Wheeling & Lake Erie**

This is the building erected around 1900 to replace the "Palmer House" station, that had stood on the same site. Until 1938, the Lake Shore Electric Railway also used the station.

Howard W. Ameling Collection

Fremont, Ohio **Wheeling & Lake Erie**

Some of the townspeople pose for this 1920s trackside view of the station, seen in the previous photograph. At right, horses also pose with their wagon at the freight house, beside which is an old-fashioned boxcar. The station was demolished in the 1950s and its space used for a parking lot.

Howard W. Ameling Collection

Toledo, Ohio **Wheeling & Lake Erie**

The scene at the Cherry Street Station as it was on May 31, 1932 with M101, M102 and baggage car loading for a last Toledo to Zanesville, Ohio run. The street side, opposite to this view, was on a higher level than the station with a short bridge walk connecting the street to the station entry main door. The station was of a Victorian design of brick construction and featuring a hipped and gabled multi-leveled roof with ornamental ridging over all the roof peak lines. Gustave Erhardt.

Station Index